CW01370853

THE BRAIN

Exploring the Latest Research and
Neurotechnological Innovations in Cognition
and Consciousness

DISCLAIMER

The author made every effort to ensure the accuracy and reliability of the information provided in this book, but it is ultimately the responsibility of the reader to exercise discretion and judgment when applying the contents herein.

Furthermore, the author hereby absolves themselves of any liability for the outcomes that may stem from the use, application, or reliance upon the information provided within this book. This disclaimer encompasses, but is not limited to, any potential loss, injury, or damage – whether direct, indirect, incidental, consequential, or punitive – resulting from actions taken based on the information herein.

By accessing and utilizing the contents of this book, readers acknowledge and agree to indemnify and hold harmless the author from any and all claims, demands, or liabilities that may arise.

COPYRIGHT STATEMENT

Copyright © Sophie Domingues-Montanari, 2024.

All rights reserved. No part of this publication may be reproduced, distributed, or transmitted in any form or by any means, including photocopying, recording, or other electronic or mechanical methods, without the prior written permission of the publisher, except in the case of brief quotations embodied in critical reviews and certain other noncommercial uses permitted by copyright law.

This book is protected under international copyright laws and treaties. Unauthorized reproduction, distribution, or transmission of this work is prohibited and may result in severe civil and criminal penalties.

THE BRAIN

Exploring the Latest Research and Neurotechnological Innovations in Cognition and Consciousness

SOPHIE DOMINGUES-MONTANARI, PhD

TABLE OF CONTENTS

Introduction: Discovering the Mysteries of the Human Brain   15

The Human Brain: Marvel of Biological Engineering   19

The Complexity of the Human Brain   21
- ❖ Introduction to the Human Brain   21
- ❖ Brain Structure   23
- ❖ Cognitive Functions   25
- ❖ Motor Control   28
- ❖ Brain and Emotions   31

The Unexplored Potential of the Brain   35
- ❖ History of Brain Research   35
- ❖ Recent Advances   38
- ❖ Unanswered Questions   41
- ❖ Enigmatic Cases and Intriguing Experiences   47
- ❖ Challenges and Limitations of Brain Research   55

Brain Wellness: Maximizing Brain Health and Longevity   59

Role of Nutrition in Brain Health   61
- ❖ Impact of Nutrition on the Brain   61
- ❖ Nutrition and Brain Health Disorders   65
- ❖ Essential Nutrients   68
- ❖ Specific Diets for Cognition   69
- ❖ Gut Microbiota, Diet, and Brain   71
- ❖ Biotechnology for Enhancing Nutrition   74
- ❖ Future Perspectives   77

Biotechnology for Brain Longevity   81
- ❖ Biotechnology for Brain Longevity   81
- ❖ Brain Preservation and Regeneration   82

❖	Genetic Factors and Brain Longevity	86
❖	Prevention of Neurodegenerative Diseases	88
❖	Ethical Challenges of Brain Longevity	90
❖	Future Perspectives	92

Optimizing the Brain: Strategies to Enhance Cognition   95

Neuroplasticity for Brain Remodeling   *97*

❖	Understanding Neuroplasticity	97
❖	Enhancement of Neuroplasticity	102
❖	Exceptional Human Performance	107
❖	Limits and Precautions	113
❖	Future Perspectives	115

Enhancing Cognition through Genetic Editing   *117*

❖	Genetic Editing and Brain Modification	117
❖	Possibilities of Cognitive Modification	121
❖	Medical and Therapeutic Applications	126
❖	Ethical Dilemmas	128
❖	Regulatory and Normative Framework	130
❖	Future Perspectives	131

Into the Inner Infinite: The Mysteries of Consciousness   135

Exploring the Human Consciousness   *137*

❖	Nature of Consciousness	137
❖	Neurobiological Foundations	140
❖	Theories of Consciousness	144
❖	Self-Awareness and Perception	151
❖	Alterations of Consciousness	154
❖	Artificial Consciousness	158
❖	Collective Consciousness	160

Redefining Consciousness through Quantum Physics   *165*

❖	Quantum Concepts Applied to Consciousness	165

- ❖ Challenges in Theoretical and Experimental Research 175
- ❖ Links Between Consciousness and the Universe 177
- ❖ Practical Applications and Implications 180
- ❖ Ethical Debates 182
- ❖ Future Perspectives 183

Digital and Spiritual Synergy: Elevating Consciousness through Technological and Spiritual Integration **187**

Spiritual Biohacking: Fusion of Technology and Spirituality 189
- ❖ Convergence between Technology and Spirituality 189
- ❖ Technologies of Spiritual Biohacking 191
- ❖ Ethics and Spiritual Challenges of Biohacking 196
- ❖ Perspectives 198

Augmented Reality for Elevating the Mind 201
- ❖ Evolution of Augmented Reality 201
- ❖ Impact on Perception and Cognition 203
- ❖ Challenges of Augmented Reality 207
- ❖ Future Perspectives 209

Beyond Mental Boundaries: Advances in Brain-Machine Connection **211**

Technological Telepathy: When Thought Becomes Universal Language 213
- ❖ Brain-Computer Communication 213
- ❖ Universal Language and Mental Connection 217
- ❖ Practical Applications 220
- ❖ Ethics of Technological Telepathy 224
- ❖ Future Perspectives 226

Downloading Consciousness and Brain Immortality 229
- ❖ Possibilities of Consciousness Transfer 229

❖	Current Scientific Realities	232
❖	Elimination of Bodily Needs	233
❖	Interactions Between Consciousnesses	241
❖	Philosophical Debates	243
❖	Ethical Challenges	245
❖	Future Perspectives	247

Conclusion: Towards Inner Infinity — **249**

To Go Further — **253**

Introduction: Discovering the Mysteries of the Human Brain

The human brain, a true masterpiece of biological engineering, remains a singularly captivating domain within the realms of science and human understanding. It constitutes a dynamic and complex universe, harboring potentials yet untapped that beckon us to push the boundaries of our comprehension. More than just an organ, the brain stands as the seat of thought, consciousness, and the architect of every aspect of our existence.

This exciting exploration guides us beyond the conventional limits of neurology to delve into the depths of cerebral well-being, aiming to optimize the health and longevity of this fascinating organ. Through the forthcoming pages, we will investigate the interrelations between diet and brain health, immerse ourselves in the advancements of biotechnology to promote brain longevity, and unveil strategies aimed at maximizing cognition.

We will also venture into the realm of neuroplasticity, this remarkable ability of the brain to reshape itself. Genetic editing, an emerging frontier, will also be scrutinized to apprehend its potential for enhancing our cognition.

The mysteries surrounding human consciousness will also form the epicenter of our journey. A thorough exploration will lead us to examine the most enigmatic aspects defining our conscious existence. From the exploration of human consciousness to a bold reinterpretation through quantum physics, our journey will traverse stimulating intellectual territories.

This book goes beyond a one-dimensional exploration of the brain. It adopts a holistic approach, seeking synergy between the digital and the spiritual to elevate consciousness. Spiritual biohacking, an innovative fusion of technology and spirituality, will be examined, as will the impact of augmented reality on our minds.

Finally, we will immerse ourselves in the remarkable progress of brain-machine connection. From technological telepathy to the bold prospect of consciousness uploading and cerebral immortality, we will explore horizons where the distinction between man and machine fades.

This work, intertwining science and speculation, prompts the reader to reevaluate their conception of the brain and consciousness. At the intersection of sciences, emerging technologies, and philosophy, this exploration seeks to illuminate the dark areas of our understanding, paving the way for profound reflections on our own existence.

1

The Human Brain: Marvel of Biological Engineering

The Complexity of the Human Brain

This chapter explores the remarkable complexity and diversity of brain functions, highlighting the central role of this organ in regulating the cognitive, emotional, and motor processes that define our experience as human beings.

❖ Introduction to the Human Brain

The human brain, often referred to as a masterpiece of nature, is the central organ of the nervous system and plays a crucial role in understanding the human mind. This complex mass of nerve tissue, housed within the skull, is the seat of thought, emotions, memory, perception, and many other functions essential to daily life. Understanding how the brain functions is of fundamental importance to neuroscientists, psychologists, and researchers in cognitive science, as it allows us to grasp the mechanisms underlying our behavior, thoughts, and experiences.

Importance of the Brain for the Human Mind

Center of Command and Control

The brain acts as the command and control center of the human body. It regulates vital functions such as breathing, blood circulation, digestion, while also coordinating more complex activities such as walking, speech, and problem-solving. This centralization of vital and complex functions underscores the crucial importance of the brain in the survival and optimal functioning of the organism.

Seat of Thought and Emotion

The brain is also the seat of thought, consciousness, and emotions. It enables us to think critically, make decisions, experience joy, sadness, love, and a range of other emotions. Understanding how the brain generates these complex mental processes is essential for grasping human nature in all its richness.

Memory and Learning

The brain is the reservoir of memory, where information is stored and retrieved. It is also the engine of learning, facilitating the acquisition of new skills and knowledge throughout life. The brain's ability to reorganize and form new connections underscores its capacity to adapt and evolve.

Interaction with the Environment

The brain constantly interacts with the environment, processing sensory stimuli and orchestrating appropriate responses. Visual, auditory, tactile perception, and other forms of perception are all brain functions that enable us to understand the world around us.

Diversity of Brain Functions

Brain Structure

The structure of the brain is extraordinarily complex. Divided into several parts, including the forebrain, midbrain, and hindbrain, each of these regions has specific functions. The cerebral cortex, the outer layer of the brain, is particularly important in higher cognitive processes such as thinking, planning, and language.

Neural Network

The complexity of the brain lies in its network of neurons, specialized cells that transmit information in the form of electrical and chemical impulses. These neuronal connections form networks that underlie all brain functions. Understanding how these networks interact is essential for grasping the complexity of mental processes.

Cerebral Hemispheres

The two cerebral hemispheres, left and right, are specialized in different functions. The left side is often associated with logic, language, and analytical skills, while the right side is more related to creativity, intuition, and spatial perception. Cooperation between these two hemispheres contributes to the diversity of human cognitive abilities.

Brain Plasticity

Brain plasticity allows the brain to adapt to new experiences and changing environments. This ability to reshape its connections in response to learning or recovery after injury underscores the remarkable flexibility and adaptability of the human brain.

❖ Brain Structure

The brain is a complex and fascinating anatomical structure. Understanding its general anatomy and neural network is essential for grasping the physical basis of perception, memory, and decision-making.

General Anatomy

Brain, Cerebellum, Brainstem

The human brain can be divided into three main regions: the cerebrum, cerebellum, and brainstem. The cerebrum, the largest part, is responsible for many higher cognitive functions, such as thinking, memory, and language. The cerebellum, located at the back of the brain, is involved in coordinating movements and maintaining balance. The brainstem, connecting the brain to the spinal cord, controls vital functions such as breathing and heart rate.

Cerebral Hemispheres and Lobes

The two cerebral hemispheres, right and left, are separated by a structure called the corpus callosum. Each hemisphere is subdivided into four lobes: frontal, parietal, temporal, and occipital. Each lobe has specific functions. The frontal lobe is involved in planning, problem-solving, and voluntary movement control. The parietal lobe processes sensory information, while the temporal lobe is essential for hearing and memory. Finally, the occipital lobe is dedicated to vision.

Neuronal Network

Neurons and Synapses

The brain's nervous tissue is composed of basic cells called neurons. These specialized cells are responsible for transmitting information throughout the brain. Each neuron has a cell body, dendrites that receive signals, and an axon that transmits signals to other neurons. Neurons communicate with each other at junctions called synapses, which are junctions where signals are transmitted by chemical substances called neurotransmitters. This complex communication between neurons is essential for all brain functions.

Brain Plasticity

Brain plasticity, often described as the brain's ability to remodel itself in response to experience, is a crucial aspect of its functioning. This plasticity manifests at different levels, ranging from the formation of new synaptic connections to the reorganization of entire brain regions. Brain plasticity is particularly evident during development, where the brain adapts to new information and experiences. However, it persists throughout life, allowing individuals to adapt to new challenges, learn new skills, and recover after brain injury.

Brain plasticity is closely linked to the brain's ability to form new synapses or adjust the strength of existing connections. Learning and memory are processes that heavily depend on this plasticity. For example, when we learn something new, physical changes occur at the synaptic level, strengthening or weakening connections between neurons. This remarkable adaptability of the brain underscores its ability to evolve based on the environment and individual experiences.

❖ Cognitive Functions

The human brain enables sensory perception, memory, thinking, and reasoning. These interconnected processes form the very fabric of our daily experience, shaping how we understand the world around us, learn, and make decisions.

Sensory Perception

Vision, Hearing, Touch, Taste, Smell

Sensory perception is the gateway to our interaction with the external world. Vision, with its millions of photoreceptors in the retina, allows us to perceive light and form complex images. Hearing, through the transformation of sound waves into electrical signals, provides us with the ability to hear and interpret various sounds. Touch, involving skin receptors, gives us the sensation of texture, pressure, and heat. Taste and smell, closely related, are chemical senses that allow us to perceive and appreciate a wide range of flavors and odors. These senses work together to create a holistic representation of our environment.

Each of these senses is processed in specific regions of the brain. Vision is primarily processed in the visual cortex, hearing in the auditory cortex, and so on. The integration of these sensory inputs occurs at the associative cortex level, where past experiences, emotions, and contexts influence our current perception.

Memory

Types of Memory: Short-Term, Long-Term

Memory is an essential aspect of cognition, allowing for the storage and retrieval of information. Generally, two types of memory are distinguished: short-term and long-term. Short-term memory is responsible for the temporary storage of information, often limited to a few seconds. This is what allows us to remember a phone number just long enough to dial it. Long-term memory, on the other hand, involves storing information over an extended period, ranging from a few hours to a lifetime. These two types of memory work together to enable us to function smoothly in our daily lives.

Mechanisms of Formation and Recall

Memory formation involves several complex processes. Encoding is the process by which sensory information is transformed into a form that can be stored in the brain. The storage of this information occurs in different brain regions, depending on the type of memory. Recall is the process of retrieving this stored information.

The mechanisms of formation and recall are influenced by factors such as attention, repetition, emotion, and context. Particularly emotional memories, for example, tend to be better encoded and recalled. Repetition also strengthens memory formation, creating more robust neuronal connections.

Thought and Reasoning

Cognitive Processes

Thought and reasoning are higher cognitive processes that distinguish humans from other species. Thinking involves the mental manipulation of information, the generation of ideas, and problem-solving. Reasoning processes allow for drawing conclusions from given information. These complex cognitive abilities involve the coordination of various brain regions, including the prefrontal cortex, the seat of executive functions.

Abstract thinking, planning, creativity, and problem-solving are examples of cognitive processes. These activities rely on memory, attention, and executive skills to produce coherent and adaptive outcomes.

Decision-Making

Decision-making is a fundamental aspect of human thinking. It relies on evaluating different options, foreseeing potential consequences, and selecting the best alternative. Emotions often play a role in this process, influencing our preferences and choices.

The neurobiology of decision-making involves brain structures such as the prefrontal cortex, the striatum, and the amygdala. These regions interact to assess risks, assign values to options, and regulate emotional responses that may influence the decision.

❖ Motor Control

Motor control, essential for interacting with our environment, is one of the most complex and intriguing aspects of human brain function. This process involves the coordination of multiple systems, from the brain to muscles, to generate precise and adaptive movements.

Motor System and Movement

The motor system encompasses all anatomical structures and physiological processes that contribute to movement production. It consists of several levels, from the brain to peripheral muscles, and operates integratively to enable a wide range of movements, from simple actions like grasping an object to more complex movements such as walking or dancing.

The Brain

The brain plays a central role in motor control. Motor areas, primarily located in the frontal lobe, are the main actors in planning, initiating, and regulating movements. The primary motor area, also called the primary motor cortex, is particularly important. It is responsible for executing voluntary movements, with each body region represented somatotopically, meaning different parts of the body are assigned to specific areas of this cortex region.

The Spinal Cord

The spinal cord acts as an essential interface between the brain and muscles. It contains neuronal circuits that can generate simple motor responses without requiring direct brain command. These reflexes are vital for quick and adaptive responses to environmental stimuli.

The Peripheral Nervous System

The peripheral nervous system transmits motor signals from the spinal cord to muscles and effector organs. Motor nerves carry signals from the brain to skeletal muscles, thereby enabling voluntary movement. Muscles, in turn, convert these nerve signals into physical movements.

Muscles

Muscles, composed of muscle fibers, are the final executors of movement. When they receive nerve signals, they contract, generating force that moves bones and joints, thereby producing movement.

Involvement of Motor Areas of the Brain

Primary Motor Area

The primary motor area, located in the precentral gyrus of the frontal lobe, is crucial for generating voluntary movements. Each specific part of this region controls a specific part of the body. Neurons in this area send motor signals to muscles, thereby triggering desired movements. The somatotopic representation of the primary motor area means that the size of the region corresponds to the amount of motor control needed for a given body part, with hands and face occupying a disproportionate share, highlighting their importance in complex movements.

Supplementary Motor Area

The supplementary motor area, also located in the frontal lobe, is involved in planning and coordinating complex movements involving multiple body parts. It plays an essential role in sequencing and coordinating movements, particularly in activities such as dancing or handling complex objects.

Prefrontal Cortex

The prefrontal cortex, although often associated with more complex executive functions, also plays a role in motor control. It is involved in long-term planning, decision-making, and inhibition of undesirable movements. Connections between the prefrontal cortex and motor areas contribute to coordinating movements based on individual goals and intentions.

The Cerebellum

The cerebellum, located at the back of the brain, is crucial for motor coordination and movement precision. It receives sensory information about body and limb positions and compares them to brain motor commands. This feedback allows the cerebellum to adjust and correct movements in real-time, ensuring precise execution of motor tasks.

Basal Ganglia

The basal ganglia, located in subcortical regions, contribute to regulating voluntary movements. They act by filtering and modulating signals from the motor cortex, thus influencing the selection and initiation of movements. They are also involved in motor planning and learning.

❖ Brain and Emotions

The human brain is closely linked to emotional processes. Emotions, subjective experiences that can range from happiness to sadness, from fear to love, are profound manifestations of our inner life.

Emotional Connections

The Limbic System

The limbic system, often considered the emotional center of the brain, is a series of interconnected brain structures. The amygdala, a key component of the limbic system, is involved in processing emotions, particularly in response to stimuli perceived as threatening or rewarding. The hippocampus, also located in the limbic system, is crucial for emotional memory, helping to store and retrieve memories related to emotional experiences.

The Prefrontal Cortex

The prefrontal cortex, especially the orbitofrontal cortex, plays an essential role in emotional regulation. It is involved in modulating responses, decision-making, and understanding the emotional consequences of actions. Connections between the prefrontal cortex and the limbic system allow for cognitive control, aiding in regulating emotions adaptively.

The Thalamus

The thalamus, acting as a kind of sensory relay, transmits sensory information to the cortex and limbic system. It plays a key role in perceiving emotional stimuli, relaying signals that trigger emotional responses. The thalamus's ability to filter and prioritize sensory information contributes to the selective aspect of our emotional responses.

Mirror Neurons

Mirror neurons, located in the motor cortex, are also involved in emotional connections. These neurons not only activate during the execution of an action but also when observing that action in others. This ability to feel what others feel, or empathy, plays a crucial role in our understanding and emotional connection with others.

Impact on Mental Well-being

Influence on Behavior

Emotional connections have a profound influence on our daily behavior. Emotions act as signals, guiding our actions and reactions. For example, fear can trigger a flight response to a perceived threat, while joy can motivate positive social behaviors. Understanding these emotional connections can be crucial for promoting adaptive behaviors and stress management.

Relationship with Memory

Emotions are closely linked to memory. Emotionally charged events tend to be better remembered than neutral events. The emotional impact on memory can have significant implications for mental well-being as it can influence how we perceive and interpret past experiences.

Stress and Emotional Response

Stress, an emotional response to perceived challenges, has significant repercussions on mental well-being. The stress response system, including hormones such as cortisol, can have lasting effects on the brain and body. Chronic stress can contribute to mental health problems.

Role in Emotional Disorders

Emotional disorders, such as anxiety and depression, are often associated with alterations in the brain's emotional connections. Imbalances in neurotransmitters, structural changes in the brain, and dysfunctional emotional responses can all contribute to these disorders. Understanding these mechanisms is essential for developing effective therapeutic interventions.

Importance of Emotional Regulation

Emotional regulation, or the ability to moderate and adjust emotional responses, is crucial for mental well-being. Effective emotional regulation enables coping with life's challenges, maintaining healthy relationships, and fostering resilience to stress. Techniques such as meditation, mindfulness, and cognitive-behavioral therapy are often used to strengthen this skill.

The Unexplored Potential of the Brain

This chapter describes recent advances in brain research and identifies unexplored areas of the human brain, shedding light on latent abilities that could redefine our perception of intelligence and consciousness.

❖ History of Brain Research

The exploration of the brain, long considered one of the most mysterious organs of the human body, has evolved through different eras, reflecting the advancement of scientific knowledge and research methods.

Ancient Times and Philosophical Foundations

The earliest inquiries into the nature of the brain date back to antiquity, where knowledge was often based on philosophy and speculation rather than scientific observation. In ancient Greece, thinkers such as Hippocrates (460-377 BC) regarded the brain as the seat of intelligence, but the exact understanding of its functioning remained largely theoretical.

However, the renowned philosopher Aristotle (384-322 BC) proposed a different perspective. He believed that the heart was the center of thought and that the brain primarily served to cool the blood. This idea, though incorrect, persisted for centuries, illustrating the resistance to paradigm shifts in the field of brain research.

Renaissance and Early Brain Anatomy

The Renaissance marked a period of rediscovery of ancient knowledge and new scientific explorations. Andreas Vesalius

(1514-1564), a Flemish anatomist, played a crucial role in challenging some preconceived ideas. In his work "De humani corporis fabrica" (1543), Vesalius presented detailed anatomical illustrations, including those of the brain, contributing to a better understanding of the structure of the nervous system.

Pioneers of Brain Physiology

In the 17th century, brain research began to transition towards a more experimental approach. Thomas Willis (1621-1675), an English physician, was among the first to suggest that the brain was the center for control and coordination of the body. His work laid the groundwork for brain physiology and introduced concepts such as cerebral circulation.

In the 18th century, advancements in the understanding of chemistry and physiology opened new perspectives. Albrecht von Haller (1708-1777), a Swiss physiologist, contributed to the understanding of nerves and introduced the concept of excitability of nervous tissues. These ideas paved the way for future research on the electrical functions of the brain.

Electricity and Reflexes: Birth of Neurophysiology

The 19th century witnessed major breakthroughs in understanding the brain, partly driven by progress in the field of electricity. Luigi Galvani (1737-1798) and Alessandro Volta (1745-1827) paved the way for studying electrical processes in the body. This electrical exploration led to significant discoveries, including the work of Emil du Bois-Reymond (1818-1896) on nerve action potentials.

Claude Bernard (1813-1878), a French physiologist, deepened our understanding of the role of the nervous system in regulating bodily homeostasis. His experiments on blood sugar regulation laid the groundwork for understanding the autonomic nervous system.

Brain Mapping and Neuroanatomy

The late 19th and early 20th centuries saw significant advances in neuroanatomy. Santiago Ramón y Cajal (1852-1934), considered the father of modern neuroscience, used silver staining to identify neurons and describe their structure. His work established that the nervous system is composed of distinct cellular units and laid the groundwork for the neuron theory.

Brain research reached new heights in the early 20th century with the advent of brain mapping techniques. Korbinian Brodmann (1868-1918) created the famous map of the human brain, dividing the cortex into numbered regions based on cellular structure. This approach paved the way for a more precise understanding of regional brain function.

Neurochemistry and Neuroimaging

The second half of the 20th century saw a rapid expansion of brain research techniques. Discoveries in neurochemistry, including the identification of neurotransmitters such as serotonin and dopamine, laid the groundwork for understanding the molecular basis of brain functions.

The advent of neuroimaging, with techniques such as positron emission tomography (PET) and magnetic resonance imaging (MRI), revolutionized brain research by enabling direct observation of brain activity. These technologies allowed exploration of correlations between brain activity and mental functions, opening new perspectives on neurological and psychiatric disorders.

❖ Recent Advances

Brain research has experienced spectacular advancements in recent decades, propelled by the emergence of cutting-edge technologies that allow us to probe the brain in more detailed and precise ways than ever before. Among these advancements, brain imaging and neurotechnologies play a central role, opening new perspectives on understanding complex brain processes.

Brain Imaging

MRI (Magnetic Resonance Imaging)

MRI has become a cornerstone of neuroscience research. It uses magnetic fields and radio waves to generate detailed images of the brain's anatomical and functional structures. Recent advances in MRI, such as functional MRI (fMRI), allow real-time observation of brain activity by measuring changes in blood flow. This provides valuable insights into brain regions associated with specific functions and has revolutionized our understanding of brain connectivity.

PET (Positron Emission Tomography)

PET is a functional imaging technique that measures the metabolic activity of the brain using radioactive tracers. It allows observation of changes in glucose and oxygen levels, providing crucial information about metabolic processes associated with neuronal activity. Combining PET with other techniques, such as MRI, offers a more comprehensive view of brain activity, opening perspectives for understanding neurological and psychiatric disorders.

EEG (Electroencephalography)

EEG records the brain's electrical fluctuations on the surface of the skull using electrodes. Recent advancements in EEG technologies, such as dry caps and signal processing algorithms, improve recording quality and enable greater mobility. EEG offers exceptional temporal resolution, allowing real-time observation of electrical changes. This technique is crucial for studying brain rhythms, wakefulness, mental attention (vigilance), and has practical applications in sleep research and brain-computer interfaces (BCIs).

Emerging Technologies

Brain-Computer Interfaces

BCIs represent an exciting frontier in brain research. These interfaces enable direct communication between the brain and an external device, often a computer. Recent advances in BCIs have led to impressive achievements, such as thought-controlled cursors or even control of robotic prostheses. Researchers are also exploring BCIs for medical applications, including restoring mobility in paralyzed individuals.

Neurological Modulation

Techniques such as transcranial magnetic stimulation (TMS) and deep brain stimulation (DBS) allow modulation of brain activity. These non-invasive approaches offer therapeutic possibilities for treating neurological disorders such as depression, Parkinson's disease, and autism spectrum disorders.

Optogenetics

Optogenetics, a technique that uses light to control neuronal activity, offers a means to selectively manipulate specific neuronal populations. This approach has the potential to reveal the underlying mechanisms of complex behaviors and explore opportunities to modulate brain activity for therapeutic purposes.

Artificial Intelligence (AI)

AI also plays an increasingly significant role in understanding the brain. Machine learning models can analyze vast datasets of brain data to identify complex patterns and deduce relationships that may escape conventional analyses. These computational approaches help decipher the mysteries of cognition and predict brain responses.

Other Recent Progress

Neuroplasticity

Neuroplasticity, or the brain's ability to reorganize and adapt, is at the heart of much recent research. Advances in brain imaging techniques have allowed the observation of neuroplasticity in action, including in contexts such as learning, recovery after brain injury, and rehabilitation. Brain stimulation-based interventions, such as TMS and DBS, are also used to modulate brain plasticity and enhance cognitive performance.

Connectomics

Connectomics, enabling the comprehensive mapping of the brain's neuronal connections, represents a major advance in understanding brain connectivity. Techniques such as MRI and PET allow tracking neuronal pathways with increased precision. This revolutionary approach provides unprecedented insights into how brain regions interact.

Devices for Monitoring Brain Health

Portable Monitoring Devices

Portable monitoring devices, such as wearable EEG headsets and neurology headbands, allow for continuous tracking of brain activity. They provide real-time data on brainwaves, electrical signals, and other parameters, thus offering valuable insights for early detection of neurological issues.

Implantable Sensors and Biocensors

Implantable sensors and biocensors offer a more invasive but highly precise approach to monitoring brain health. These devices can be implanted directly into the brain to monitor parameters such as intracranial pressure, temperature, and neurotransmitter levels. They are particularly useful in managing severe neurological conditions.

❖ Unanswered Questions

Recent technological developments offer new ways to probe the complexities of the brain. However, despite these advances, many mysteries remain, challenging researchers and fueling continued fascination with the human brain.

Brain Mapping: Current Knowledge and Grey Areas

Brain mapping is an ambitious task undertaken by researchers worldwide. It involves identifying and detailing brain structures as well as their complex interactions. Brain regions are associated with specific functions, and mapping helps understand how these areas interact to support cognition, emotions, and movement.

Several advanced techniques are employed in brain mapping, but MRI and PET are among the most commonly used. MRI provides detailed images of brain structure, while PET offers insights into neuronal activity.

Many brain structures and functions have already been precisely identified and mapped. These include:

- Cerebral Lobes: The brain's four main lobes - frontal, parietal, temporal, and occipital - have been mapped to understand their specific roles. For example, the frontal lobe is often associated with executive functions and motor control, while the temporal lobe is involved in auditory processing and memory.
- Gyri and Sulci: The convolutions (gyri) and grooves (sulci) of the cortical surface have been meticulously mapped to understand brain topography. Some gyri and sulci are associated with specific functions, and their mapping helps locate precise brain areas.
- Language Centers: Brain mapping has identified specific areas associated with language production and comprehension. Broca's area, for example, is involved in language production, while Wernicke's area is related to comprehension.
- Motor and Sensory Cortex: Regions of the motor and sensory cortex have been mapped to understand the

spatial representation of the body in the brain, often referred to as the cortical homunculus. This helps locate areas responsible for movement and sensory perception for different body parts.
- Hippocampus: This key structure in the temporal lobe is crucial for spatial memory and has been mapped to understand its role in memory formation.
- Thalamus and Hypothalamus: These subcortical structures have been mapped to understand their role in regulating vital functions, sleep, and sensory signal transmission.

These advancements have significant implications in neurology, psychology, and psychiatry, providing insights into understanding and treating conditions such as Alzheimer's disease, depression, and schizophrenia.

However, grey areas persist.

- One major grey area lies in precisely understanding the mechanisms underlying consciousness. Although brain mapping has identified brain regions associated with specific functions, the exact nature of consciousness remains an enigma.
- Furthermore, interindividual variability poses a significant challenge. As each brain is unique, brain mapping must account for this diversity to develop a more precise understanding of individual differences in brain structure and function. Responses to stimuli, perception thresholds, and even anatomical layout can vary considerably from person to person.
- The complex interactions between different brain regions represent another obscure area. Understanding how these regions work together to generate thoughts, emotions, and movements requires fine mapping of

neuronal connections and functional networks, a complex task still under development.

Understudied Brain Regions

Certain areas of the brain, although crucial, remain understudied due to their complexity or difficulty of access. The amygdala complex, for example, located deep in the brain, is involved in emotion processing, but the precise mechanisms of its function remain a mystery. Similarly, the dentate gyrus, a region of the hippocampus, is associated with the formation of new memories, but its exact role remains unclear.

These understudied regions hold untapped potential that could offer innovative perspectives on human cognition. Exploring these areas could reveal unexpected mechanisms and subtle interactions that contribute to crucial aspects of thought, perception, and decision-making.

In this context, emerging technologies, such as optogenetics, which allows precise manipulation of neurons using light, open new avenues to probe these once inaccessible areas.

Unsolved Mysteries

Nature of Consciousness

The question of consciousness remains among the most enigmatic. How does it emerge from neuronal activity? What gives rise to our subjective experience of the world? These questions remain unanswered.

David Chalmers introduced the term "hard problem of consciousness" to describe this philosophical challenge related to explaining the subjective nature of conscious experience, also known as qualia. He distinguishes between the "easy problem" of consciousness, which concerns understanding the brain mechanisms related to cognitive functions, and the "hard problem," which explores why and how these brain processes give rise to a subjective experience, an aspect that seems difficult to explain solely through a neuroscience or cognitive approach. This distinction has significantly influenced philosophical and scientific debates on the nature of consciousness.

Emerging technologies, especially fine mapping of neuronal connections and precise modulation of brain activity, could offer new insights into the underlying mechanisms of human consciousness.

Memory Mechanisms

Although we understand some aspects of memory, such as short-term and long-term storage, the precise mechanisms of memory formation, consolidation, and retrieval continue to elude complete understanding.

Understanding Emotions

Emotions, essential to our human experience, are not fully deciphered. The exact neurobiological mechanisms underlying the complex range of human emotions remain largely unexplored.

Creativity

Creativity, the ability to generate new and original ideas, is one of the most fascinating abilities of the human brain. The brain processes underlying creativity largely escape linear explanation. Areas such as the prefrontal cortex, anterior cingulate cortex, and striatum are associated with specific aspects of creativity, but how these regions interact to give rise to creative works remains largely a mystery.

Intuition

Intuition, the ability to understand or perceive something without resorting to conscious analysis, is another intriguing dimension of brain potential. Some studies suggest that intuition could be linked to subconscious processes in regions such as the limbic system, but the precise extent of its origin remains unclear.

Emotional Resilience

Emotional resilience, the ability to cope with emotional challenges and adapt to adversity, remains an area where neurobiological mechanisms are still poorly understood. Areas such as the amygdala and prefrontal cortex seem to play a role, but how these regions interact to promote resilience remains an active area of investigation.

Perspectives on Major Advancements

Personalized Treatments for Neurological Disorders

Advancements in connectomics, neurological modulation, and ICM (Integrative Computational Models) are paving the way for more personalized treatments for neurological disorders. Understanding an individual's specific connectivity and developing targeted interventions could revolutionize how we treat conditions such as depression, epilepsy, and Alzheimer's disease.

Understanding Psychiatric Disorders

Emerging technologies allow for a deeper exploration of the neurobiological bases of psychiatric disorders such as schizophrenia, bipolar disorder, and depression. This could lead to more precise diagnoses and more effective interventions, thus improving the quality of life for affected individuals.

New Therapies for Brain Plasticity

Precise neurological modulation, combined with a better understanding of brain plasticity, offers opportunities to develop therapies aimed at stimulating recovery after brain injuries, improving cognitive performance, and treating conditions such as autism spectrum disorders.

❖ Enigmatic Cases and Intriguing Experiences

Intriguing brain phenomena that defy our conventional understanding sometimes emerge. These enigmatic cases, ranging from altered perception experiences to astonishing cognitive abilities, evoke profound interest both in the scientific community and in the collective imagination.

Examples of Individuals with Extraordinary Abilities

Shakuntala Devi

Shakuntala Devi, known as "the human computer," possessed extraordinary mathematical abilities. Born in 1929 in India, she demonstrated exceptional mental calculation skills from a young age. Her fame peaked in 1980 when she mentally calculated the product of two 13-digit numbers in 28 seconds.

Alexis Lemaire

Alexis Lemaire, born in 1980, is a Frenchman with remarkable mathematical abilities. In 2002, Lemaire calculated the 13th root of a 200-digit number in 70.2 seconds, setting a Guinness World Record. His exceptional speed and accuracy in complex mental calculations garnered admiration from the public and experts alike.

Terence Tao

Terence Tao, born in 1975 in Australia, showed exceptional mathematical aptitude from a young age. He earned his PhD at 21 from Princeton University and at 24, became a full professor at the University of California, Los Angeles (UCLA), becoming one of the youngest full professors in UCLA's history. At 31, he was awarded the Fields Medal, the highest honor in mathematics, for his outstanding work in harmonic analysis, partial differential equations, combinatorics, number theory, and additive mathematics.

William James Sidis

Born in 1898, William James Sidis was considered one of the most intelligent individuals in history. At 11, he was admitted to Harvard, becoming the youngest student ever enrolled there. A polyglot, he spoke multiple languages and wrote on various subjects. However, despite his exceptional abilities, Sidis had a difficult relationship with society, preferring to withdraw from media attention. His story underscores the challenges that exceptionally intelligent individuals may face.

Christopher Hirata

Christopher Hirata Christopher Hirata is a renowned American-Japanese physicist whose remarkable abilities shine in the fields of mathematics and theoretical physics. Born in 1982, he earned his undergraduate degree at the age of 18 and obtained his doctorate at 22 from Princeton University. His exceptional contributions include work on string theory and cosmology.

Srinivasa Ramanujan

Born in 1887 in India, Srinivasa Ramanujan developed mathematical theories without formal training. His results in areas such as infinite series, continued fractions, and prime numbers astonished the mathematical world. Despite difficult living conditions, Ramanujan produced thousands of mathematical results, many of which became starting points for further research. His contributions to number theory and mathematical analysis have had a lasting impact, leading to his election as a Fellow of the Royal Society of London.

Daniel Tammet

Daniel Tammet is a British writer and autistic savant born in 1979. He is best known for his exceptional skills in memorization and mental calculation. Diagnosed with Asperger's syndrome and savant syndrome, Tammet set a record by memorizing and reciting the digits of Pi to 22,514 decimal places. Tammet is also a polyglot, having learned Icelandic in a week to participate in a television show.

Kim Peek

Born in 1951 in the United States and diagnosed with savant syndrome, Kim Peek was a prodigious memorizer with eidetic memory, capable of retaining extremely complex details. His ability to read two pages simultaneously (one with each eye) and memorize thousands of books fascinated the world. Kim Peek inspired the character of Raymond Babbitt in the 1988 film "Rain Man."

Derek Paravicini

Derek Paravicini is a British pianist born in 1979. Blind and intellectually disabled due to extreme prematurity, Paravicini developed an extraordinary talent for music from a young age. Capable of playing virtually any piece after hearing it once, he has mastered a vast repertoire of musical styles. His gift is attributed to perfect pitch and prodigious musical memory.

Intriguing Phenomena

Phineas Gage Case

One of the most famous and intriguing case studies in the history of neurology is that of Phineas Gage. In 1848, Gage survived an accident in which an iron rod pierced his skull, severely damaging his frontal lobe. To the surprise of doctors, Gage survived, but his behavior radically changed. Once steady and responsible, he became impulsive and irresponsible. This case laid the foundation for understanding the role of the frontal lobe in personality and behavior.

Mnemonic Abilities

Cases of exceptional memory raise questions about the limits of human mnemonic capacity. These individuals can recall intricate details of their daily lives over long periods, often without conscious effort.

Studying these cases has led to the discovery of neurological differences in brain structures associated with memory. However, how these differences explain the ability to remember exceptionally remains an active area of research. The implications of these findings for the overall understanding of memory and cognition are vast.

Jill Price's case in this context is remarkable. Born in 1965, she can recall nearly every day of her life since adolescence in detail. Her mind spontaneously retains past events, conversations, and even weather conditions associated with each day. This exceptional memory, while fascinating, can sometimes be overwhelming, affecting her ability to focus on the present.

Multiple Personality Disorders

Multiple personality disorders, now referred to as dissociative identity disorders, present cases where a person appears to harbor several distinct personalities, each with its own memories and characteristic traits.

While controversial and sometimes questioned, these cases highlight the complexity of consciousness and identity. Brain imaging studies have revealed changes in the activation of certain brain regions depending on the manifested personality.

However, understanding how these different personalities can coexist within the same brain remains a major challenge for neuroscience research and raises fundamental questions about the nature of consciousness and personal identity.

In this context, a well-documented example is the case of Sybil Dorsett. Dorsett was an American artist who consulted psychoanalyst Cornelia B. Wilbur in the 1950s for anxiety issues. Over the course of sessions, Wilbur discovered that Dorsett exhibited sixteen distinct personalities, each with its own name, behavior, and memory. The personalities, including children, men, and women, resulted from severe physical and sexual abuse endured during childhood.

Acquired Savants

Acquired savants are individuals who, after a brain injury or illness, suddenly develop extraordinary artistic, mathematical, or musical talents without prior training.

These cases raise questions about latent reservoirs of skills in the human brain that can be triggered by specific circumstances.

Understanding the mechanisms of this brain plasticity could lead to innovative therapeutic interventions to enhance cognitive performance or promote recovery after brain injuries.

Here are some examples of acquired savants:

- Orlando Serrell: After being hit in the head by a baseball at the age of 10, Serrell developed the exceptional ability to memorize the days of the week, dates, and weather conditions of each day since the accident.

- Alonzo Clemons: After suffering a brain injury from an accident, Clemons developed an extraordinary talent for sculpting lifelike animals from clay, despite a mild intellectual disability.

- Jason Padgett: After an assault, Padgett, who previously had no interest in mathematics, developed an exceptional ability to visualize complex mathematical formulas and create artwork based on these concepts.

- Tony Cicoria: After surviving a lightning strike, Cicoria developed a passion for music and became a talented composer, despite his previous lack of interest in music.

- Derek Amato: Derek Amato became a talented pianist after a concussion from diving into a pool. Developing a "musical savant," he acquired exceptional ability to play the piano without prior training, illustrating the remarkable plasticity of the human brain in response to trauma.

Surprising Experiences

Synesthesia

Synesthesia is a fascinating phenomenon where sensory stimuli cross, causing a simultaneous experience of multiple senses. For example, a synesthete may perceive colors while listening to music or involuntarily attribute flavors to words. This mysterious phenomenon, although long considered rare, is more common than expected, affecting about 1 in 200 people. Researchers have identified neurological correlates of synesthesia, suggesting that unusual connections between brain regions responsible for sensory perception may underlie these experiences.

Altered States of Consciousness

Altered states of consciousness refer to temporary alterations in perception, cognition, and subjective experience that differ from the ordinary state of consciousness. These states can be induced by various factors such as meditation, psychoactive substances, fasting, sleep deprivation, or specific medical conditions like schizophrenia. Altered states of consciousness can vary in intensity and nature, ranging from subtle changes in perception such as visions, to profoundly altering transcendental experiences of reality.

Researchers are interested in altered states of consciousness (ASC) to understand the nature of consciousness and subjective experience. Indeed, these states allow for exploring the limits of human perception, the underlying mechanisms of consciousness, and the ways in which the brain generates subjective reality. Brain imaging has enabled the observation of neurological changes associated with these states, but the understanding of the precise mechanisms involved remains limited. Some researchers suggest that these phenomena could stem from the desynchronization of normal brain networks.

Lucid Dreams

Lucid dreams, where a person becomes aware that they are dreaming and can often exert some control over the dream's course, are another intriguing case of inexplicable brain phenomena. Although lucid dreams are increasingly studied, their understanding remains incomplete. Brain imaging studies have identified distinct patterns of brain activity during lucid dreams, but the precise mechanisms of this dream consciousness remain poorly understood. Some researchers speculate on the role of brain regions associated with metacognition and decision-making in the emergence of this fascinating ability.

Near-Death Experiences

Near-death experiences (NDEs) typically occur in life-threatening situations, where individuals report experiences of bright tunnels, encounters with deceased loved ones, or life reviews. These experiences, though often profound and personal, defy a clear scientific explanation. Neuroscientists are exploring the brain mechanisms that could contribute to these experiences, but the mystery persists.

Past-Life Memories

Some children, often between three and five years old, report detailed memories of past lives, with precise details about places, events, and people. While these accounts can often be explained by other factors, they raise questions about the nature of memory and the possibility of transgenerational influences on cognition. They provide fertile ground for research on memory and time perception. These phenomena could also be explored to better understand the underlying brain mechanisms of mystical and transcendent experiences.

Out-of-Body Experiences

Out-of-body experiences, where a person seems to float outside their own body, defy rational explanations. Some patients who have survived cardiac arrests report experiences of this kind, but the exact nature of these phenomena remains a mystery. Some researchers suggest that these experiences could be linked to temporary alterations in consciousness.

❖ Challenges and Limitations of Brain Research

Brain research faces a series of challenges and limitations inherent in the complexity of the human brain. These obstacles can be technological, ethical, or conceptual.

Complexity of the Brain

The human brain consists of approximately 85 billion neurons forming a complex network interconnected by synapses. Additionally, it is estimated that about 100 billion glial cells accompany these neurons. These cells play an essential role by providing structural and nutritional support to neurons, contributing to maintaining the brain's chemical balance, and actively participating in brain immune defense.

With 10,000 trillion synapses in 1 cubic centimeter of the brain, this network functions like an information highway, propagating signals at an impressive speed of 120 m/s along the widest nerve fibers.

The attempt to map these connections and understand the underlying mechanisms of cognitive and emotional processes is thus hindered by the complexity of the neuronal network. Current technologies only allow for a partial view, and simplified models, although useful for establishing foundations, still fall short of reflecting the infinitely complex reality of neuronal activity.

Limits of Brain Imaging

Techniques such as MRI, PET, and EEG provide valuable insights into brain activity, but they face significant technological obstacles. The limited spatial and temporal resolution of these techniques hinders the ability to observe brain events on a fine scale and in real time. Moreover, some brain areas remain difficult to probe due to their inaccessibility or sensitivity to magnetic fields, limiting our understanding of specific functions.

Limits of Computer ModelingTop of Form

Reproducing the complexity of the human brain in a computer model remains a colossal challenge. The necessary simplification to make these models manageable can lead to representations that lack biological fidelity. Moreover, current models struggle to account for brain plasticity, learning, and memory in an integrated manner.

Individuality and Brain Variability

Brain structure and function can vary significantly from one person to another, making it difficult to draw general conclusions. Genetic, environmental, and developmental factors contribute to this variability, complicating the task of making generalizations from research findings.

Limits of Correlation and Causality

Observing a correlation between specific brain activity and behavior does not guarantee a direct causal relationship. The complex relationships between different parts of the brain and the multiple factors influencing human behavior complicate the determination of causality links.

Ethical Issues in Brain Research

Advancements in human brain research also raise profound ethical concerns. The ability to manipulate brain activity raises questions about mental privacy, individual autonomy, and potential risks. Brain-machine interfaces, while offering rehabilitation prospects for disabled individuals, also raise concerns about safety and ethical use of such technologies.

The use of AI in research also raises ethical questions, particularly regarding algorithm transparency, accountability, and ethical use of data. The possibility of predicting mental states or behaviors from brain data poses ethical challenges related to privacy and potential stigmatization.

Another crucial ethical issue lies in the possibility of enhancing brain capabilities beyond the norm. The temptation to resort to interventions to increase memory, creativity, or other cognitive functions raises fundamental questions about equitable access to such technologies and the social risks associated with such enhancements.

Brain research thus faces the need to develop robust ethical frameworks to guide the responsible application of these emerging technologies.

2

Brain Wellness: Maximizing Brain Health and Longevity

Role of Nutrition in Brain Health

This chapter explores the impact of our dietary choices on brain function, highlights essential nutrients for optimal cognition, and examines dietary strategies conducive to mental well-being.

❖ Impact of Nutrition on the Brain

Nutrition plays a crucial role in brain health, influencing the structure and function of the brain. Balanced nutritional choices, rich in essential nutrients, can promote mental clarity, memory, and contribute to the prevention of certain neurological diseases.

Nutrition and Brain Health

Nutrition can have significant implications for the brain, affecting not only its development and daily function but also its long-term health.

To fully understand the impact of nutrition on the brain, it is essential to closely examine specific nutritional components that play a key role. Omega-3 fatty acids, for example, are crucial for the structure of brain cell membranes and are associated with improved cognitive functions. B vitamins are also essential for brain health. Vitamin B12, for instance, is involved in DNA synthesis and nerve cell formation, while vitamin B9 is crucial for nervous system development. Another important element is vitamin E, a potent antioxidant that can help protect brain cells from oxidative damage. Additionally, minerals such as iron and zinc are necessary for optimal brain function, participating in processes such as oxygen transport and nerve signal transmission.

Besides these specific components, the overall composition of the diet plays a major role. Diets rich in fruits, vegetables, whole grains, and lean protein sources provide a variety of nutrients beneficial for the brain. These foods are rich in antioxidants, which help combat inflammation, detrimental to brain health.

Short and Long-Term Effects of Nutrition on the Brain

The impact of diet on the brain may vary from person to person due to factors such as genetic heritage, level of physical activity, and other lifestyle habits. However, general principles of balanced nutrition can benefit most individuals.

Short-Term Effects

A balanced diet can have immediate effects on cognitive performance. Consuming meals rich in complex carbohydrates, proteins, and fibers promotes a steady release of energy, supporting concentration and mental clarity throughout the day. Conversely, high levels of blood sugar, often associated with diets rich in added sugars, can lead to energy and mood fluctuations.

Long-Term Effects

In the long term, healthy dietary choices contribute to the preservation of brain function. Imbalanced diets, high in saturated fats, added sugars, and salt, can affect blood flow to the brain and contribute to issues such as strokes and dementia. Diets rich in antioxidants and specific nutrients are associated with a reduced risk of neurodegenerative diseases and healthier cognitive aging.

Impact on Brain Functions

Memory

Nutrition directly influences memory. Studies show that certain nutrients, such as omega-3 fatty acids, antioxidants, and B vitamins, promote memory consolidation and help prevent age-related cognitive decline.

Concentration and Mental Clarity

The ability to focus and maintain mental clarity also depends on nutrition. Nutrient-rich foods, especially those containing glucose from complex sources like whole grains, provide stable energy necessary for optimal brain function.

Brain Plasticity

Brain plasticity, the brain's ability to change and adapt, is influenced by diet. Research suggests that certain diets, such as the Mediterranean diet rich in fruits, vegetables, and omega-3 fatty acids, promote brain plasticity, which is crucial for learning and memory.

Neurogenesis

Neurogenesis, the process of forming new neurons, mainly occurs in the hippocampus, a brain region associated with memory and learning. Preliminary studies suggest that certain nutrients, such as polyphenols found in berries and tea, may promote neurogenesis.

Concept of "Hungry Brain"

The term "hungry brain" is used metaphorically to describe a nutritional deficiency that can negatively affect cognitive performance. When the brain does not receive the necessary nutrients, its functions can be compromised, leading to decreased attention, memory, and other cognitive abilities.

Modern dietary habits, often characterized by overconsumption of processed foods rich in added sugars and saturated fats, may contribute to nutrient deprivation. This impoverished diet may potentially contribute to mental health problems and reduced cognitive performance.

To avoid the concept of a "hungry brain," it is essential to promote healthy eating habits that provide a variety of nutrients necessary for the brain. This includes consuming fruits, vegetables, fatty fish, nuts, and other nutrient-rich foods.

Importance of Hydration

Hydration plays a crucial role in optimal brain function, influencing various aspects of brain function.

Adequate hydration maintains blood volume, ensuring an adequate supply of oxygen and nutrients to the brain. Additionally, water facilitates the transport of neurotransmitters, essential for nerve signal transmission.

Dehydration can lead to decreased concentration, short-term memory, and vigilance. Studies have shown that even mild dehydration can negatively influence cognitive performance, affecting decision-making ability and problem-solving.

Furthermore, hydration maintains the regulation of body temperature, preventing brain overheating. Prolonged dehydration can lead to headaches, dizziness, and, in extreme cases, more serious complications.

❖ Nutrition and Brain Health Disorders

Nutrition plays a crucial role in preventing neurological disorders and cerebrovascular diseases. Balanced dietary choices, rich in omega-3 fatty acids, B vitamins, antioxidants, and low in salt, can promote brain health.

Neurological Disorders

Alzheimer's Disease

Alzheimer's disease, the most common form of dementia, is closely related to lifestyle factors, including diet. A diet rich in antioxidants, omega-3 fatty acids, and B vitamins, combined with weight management and blood sugar regulation, can contribute to preventing this neurodegenerative disease.

Parkinson's Disease

Although Parkinson's disease is largely linked to genetic factors, studies suggest that specific dietary habits, such as consuming antioxidants from fruits and vegetables, may offer some support in managing symptoms and slowing the progression of the disease. Additionally, certain foods, such as those rich in omega-3 and polyphenols, have anti-inflammatory properties, and chronic inflammation is a common factor in many neurological diseases, including Parkinson's disease.

Multiple Sclerosis

Multiple sclerosis is an autoimmune disease of the central nervous system. Studies indicate that diets rich in specific nutrients, such as vitamin D and omega-3 fatty acids, may have beneficial effects on preventing this disease and modulating its course.

Cognitive Function and Depression

Certain nutrients, such as omega-3 fatty acids, vitamin B12, and zinc, play a crucial role in optimal brain function, and deficiencies in these nutrients have been associated with memory problems, concentration issues, and mood disorders, including depression.

Cerebrovascular Diseases

High Blood Pressure and Salt

High blood pressure is one of the major risk factors for cerebrovascular diseases. Reducing salt intake, combined with a balanced diet including fruits, vegetables, and low-fat dairy products, can help maintain normal blood pressure, thereby reducing the risk of stroke.

Role of Fats in Cerebrovascular Diseases

The types of fats consumed in the diet play a crucial role in vascular health. Saturated fats, found in animal-based foods and processed products, can contribute to plaque buildup in the arteries, increasing the risk of stroke. In contrast, omega-3 fatty acids, found in fatty fish and nuts, have anti-inflammatory effects and can help maintain blood vessel flexibility.

Diabetes and Blood Sugar Management

Diabetes is another major risk factor for cerebrovascular diseases. A diet rich in fiber from whole grains, vegetables, and fruits can help regulate blood sugar levels. Weight control and prevention of type 2 diabetes through a balanced diet are key strategies for reducing the risk of vascular complications.

Antioxidants and Protection of Blood Vessels

Antioxidants found in fruits and vegetables play a crucial role in preventing lipid oxidation, a process associated with atherosclerosis. Antioxidant compounds such as vitamins C and E, as well as polyphenols, can help maintain the integrity of blood vessels, thereby reducing the risk of cerebrovascular diseases.

Fiber and Stroke Risk Reduction

Diets rich in fiber have been associated with a reduced risk of stroke. Dietary fiber, found in fruits, vegetables, whole grains, and legumes, can help control weight, regulate blood pressure, and maintain healthy cholesterol levels, all crucial elements for vascular health.

Role of B Vitamins

B vitamins, particularly vitamin B6, vitamin B9 (folic acid), and vitamin B12, are involved in regulating homocysteine levels, an amino acid linked to atherosclerosis. A diet rich in foods such as whole grains, green vegetables, and nuts can help maintain adequate levels of these vitamins and reduce the risk of cerebrovascular diseases.

❖ Essential Nutrients

The brain is an energy-hungry organ, accounting for approximately 20% of the body's total energy expenditure. To maintain its optimal functioning, it relies on various essential nutrients.

Omega-3 Fatty Acids: Fundamental for Brain Health

Omega-3 fatty acids, particularly docosahexaenoic acid (DHA) and eicosapentaenoic acid (EPA), are essential components of the brain's cell membranes. They play a crucial role in nerve signal transmission, synaptic plasticity, and reducing brain inflammation. They are found in fatty fish such as salmon, mackerel, and sardines. Nuts, flaxseeds, and flaxseed oil are also options for those following a vegetarian or vegan diet.

Antioxidants: Protection Against Oxidative Stress

Antioxidants are essential for protecting the brain against oxidative stress, a process involved in brain aging and various neurological disorders. They neutralize free radicals, thereby reducing the risk of cellular damage. They are found in colorful fruits and vegetables such as berries, cherries, spinach, and broccoli. Green tea, nuts, and legumes also contribute to providing a variety of beneficial antioxidants.

Vitamins: Support for Cognitive Processes

B Vitamins for Brain Function

B vitamins, especially B6, B9 (folic acid), and B12, are essential for cognitive function. They are involved in the production of neurotransmitters such as dopamine and serotonin, as well as in regulating homocysteine, linked to vascular health. Sources of B6 include bananas and nuts. Vitamin B9 is found in leafy greens or legumes. B12 is present in animal products such as meat, fish, eggs, and in some fortified foods.

Vitamin D for Neuroprotection

Vitamin D plays a role in neuroprotection and modulation of inflammatory processes. Vitamin D deficiency has been associated with an increased risk of neurological disorders, underscoring its importance for brain health. Major dietary sources of vitamin D include fatty fish (salmon, tuna), cod liver oil, fortified dairy products, eggs, certain mushrooms, and fortified orange juice. However, sunlight exposure remains the primary natural source of vitamin D.

Minerals: Essential for Nerve Transmission

Zinc and iron are essential for nerve transmission. They are involved in neurotransmitter synthesis and regulation of neuronal responses, influencing cognition and memory. Seafood, lean red meat, pumpkin seeds, and legumes are good sources of zinc. Spinach, lentils, and lean meat contribute to iron intake.

❖ Specific Diets for Cognition

Some studies suggest that diets such as the Mediterranean diet, which emphasize healthy fats, fruits, vegetables, and fish, may help reduce the risk of neurodegenerative diseases, including Alzheimer's disease.

Mediterranean Diet: A Holistic Approach to Brain Health

The Mediterranean diet is characterized by an abundance of fruits, vegetables, whole grains, fish, nuts, and olive oil. It is distinguished by low consumption of red meat and dairy products. Key components of this diet are rich in omega-3 fatty acids, antioxidants, and vitamins.

Studies have shown that the Mediterranean diet is associated with a reduced age-related cognitive decline and a lower risk of developing Alzheimer's disease. The anti-inflammatory properties of this diet, along with its influence on vascular health, contribute to its cognitive benefits.

MIND Diet: Focus on Brain-Beneficial Foods

The MIND (Mediterranean-DASH Intervention for Neurodegenerative Delay) diet combines principles of the Mediterranean diet and the DASH (Dietary Approaches to Stop Hypertension) diet. It emphasizes foods specifically beneficial for the brain, such as berries, nuts, and leafy greens.

Studies suggest that the MIND diet may reduce the risk of neurodegenerative diseases, including Alzheimer's disease. Antioxidants present in berries, in particular, are associated with better cognitive function.

Prevention of Neurological Diseases

Diets rich in essential nutrients, such as the Mediterranean diet and the MIND diet, are associated with a reduced risk of neurodegenerative diseases because they provide a variety of nutrients beneficial for brain health, acting on preventing cognitive decline.

Thus, adopting these diets early in life can strengthen the brain's defenses and contribute to better long-term cognitive health. Early prevention through a balanced diet is a key aspect of overall brain health management.

❖ Gut Microbiota, Diet, and Brain

The connection between gut microbiota and the brain is increasingly drawing attention. Some foods, such as probiotics found in yogurts and fermented foods, can promote a healthy gut microbiota, with potential implications for mental health.

Intestinal Microbiota

The gut microbiota, also known as gut flora or microbiome, refers to the complex community of billions of microorganisms that mainly reside in the digestive tract, particularly in the colon, which is the last part of the intestine. However, microorganisms can also be present in other parts of the gastrointestinal tract, including the small intestine. These microorganisms include bacteria, viruses, fungi, and other microscopic life forms. The gut microbiota plays an essential role in various aspects of human health.

Here are some points regarding the gut microbiota:

1. Microbial diversity: The microbiota consists of a wide variety of microorganisms belonging to different strains and species, and their composition can vary from person to person.
2. Metabolic functions: Microorganisms in the microbiota participate in important metabolic processes such as the digestion of non-digestible dietary fibers, the production of vitamins (such as vitamin K and certain B vitamins), and the metabolism of bile compounds.
3. Role in the immune system: The gut microbiota plays a crucial role in the development and modulation of the immune system. It contributes to educating the immune system, helping to distinguish between pathogens and normal body components. It also produces antibacterial substances that can inhibit the growth of harmful bacteria.
4. Communication with the nervous system: The microbiota can communicate with the central nervous system (CNS) through various mechanisms, including the vagus nerve and the production of neurotransmitters.
5. Evolution over time: The microbiota undergoes changes throughout a person's life, influenced by factors such as diet, antibiotics, environment, and other lifestyle-related factors.

Role of the Microbiota in Cognition

Gut-Brain Communication

The communication between the gut and the brain, often referred to as the gut-brain axis, plays a crucial role in various aspects of health. It influences mood, behavior, and cognition, and imbalances in this axis may be linked to neurological and psychiatric disorders.

This axis involves bidirectional communication between the enteric nervous system (ENS), often called the "second brain," which is a network of neurons present in the intestine, and the central nervous system (CNS), primarily the brain, thus establishing a direct connection between the gut and the brain.

Role of the Microbiota

The microbiota can interact with the ENS via neuroendocrine axes, and these interactions can influence signaling between the gut and the brain.

Here are some key points regarding the role of the microbiota and gut-brain communication in cognition:

1. Inflammatory conditions in the gut can trigger inflammatory responses in the brain, which can impact cognition and lead to neurological disorders.

2. The gut and the brain share sensitivity to stress, and stress can influence intestinal function. This interaction can contribute to disorders such as irritable bowel syndrome, which is often associated with cognitive symptoms.

3. Microorganisms in the microbiota can metabolize certain dietary compounds to produce bioactive metabolites, such as short-chain fatty acids (SCFAs). These metabolites can impact brain function and cognition.

4. Some types of bacteria present in the microbiota are capable of synthesizing neurotransmitters, including serotonin, dopamine, and GABA. These

neurotransmitters can influence the functioning of the central nervous system.

5. The microbiota plays a role in maintaining the integrity of the intestinal barrier. Impaired intestinal barrier function can allow the passage of undesirable substances into the bloodstream, triggering an immune and inflammatory response that can affect the brain.

Microbiota Diversity for Brain Health

A balanced diet rich in fiber, fruits, and vegetables promotes diversity in the gut microbiota. This diversity is crucial for maintaining a healthy balance and positively influencing brain health.

Probiotics: Allies for the Brain

Probiotics, found in foods such as yogurt and kimchi, are beneficial microorganisms that strengthen the gut flora. Studies suggest that probiotics may have positive effects on reducing inflammation and improving mental health.

Prebiotics: Nourishing Beneficial Bacteria

Prebiotics are non-digestible dietary fibers that nourish beneficial bacteria in the microbiota. Foods rich in prebiotics, such as garlic, onions, and bananas, promote the growth of probiotic bacteria, thereby supporting brain health.

❖ Biotechnology for Enhancing Nutrition

Biotechnology has opened exciting new perspectives in the field of nutrition, particularly regarding the creation of functional foods and nutraceuticals aimed at promoting brain health and the customization of diet through genomics.

Advancements in Biotechnology for Brain Nutrition

Functional Foods

One application of biotechnology in the field of nutrition for brain health lies in the genetic manipulation of foods. For example, certain foods like milk or eggs are artificially enriched with omega-3 fatty acids, which are crucial for cognition and the prevention of neurodegenerative diseases.

Another interesting area of exploration is the genetic manipulation of probiotic strains to boost the production of neuroprotective compounds, thus enhancing communication between the gut and the brain.

Furthermore, genetic manipulation facilitates the creation of functional foods rich in antioxidants and beneficial phytochemicals for the brain. These substances help protect nerve cells against oxidative stress, a factor associated with brain aging and the development of neurodegenerative diseases.

Dietary Supplements

Dietary supplements aimed at promoting brain health are often enriched with specific nutrients such as omega-3 fatty acids, antioxidants, vitamins, and minerals that play a crucial role in optimal brain function.

Thus, through biotechnology, microorganisms such as yeasts can be genetically modified to efficiently synthesize omega-3 fatty acids. This provides a vegan and sustainable alternative to traditional sources, such as fatty fish, while ensuring optimal levels of these nutrients beneficial for the brain.

Biotechnology also enables the development of personalized dietary supplements by taking into account individual genetic variations that influence how the body metabolizes certain nutrients. This allows for a more precise approach, and individuals can benefit from products tailored to their specific needs based on their genetic profile.

Targeted Delivery of Nutrients

The blood-brain barrier, a complex physiological barrier that protects the brain from potentially harmful substances, has long posed a major challenge in designing effective strategies for delivering nutrients directly to the brain. This is where biotechnology comes into play, offering innovative solutions to bypass this barrier and enable targeted delivery.

In this context, nanoparticles are one of the most studied vectors for targeted delivery of nutrients directly to the brain. These nanoscale particles can be designed to encapsulate specific nutrients while being small enough to traverse the blood-brain barrier. They are made from various materials, such as polymers, lipids, or even metals. They can also be functionalized with specific coatings, allowing them to target particular receptors on brain cells. Thus, they facilitate the release of nutrients precisely where they are needed.

Furthermore, microorganisms such as bacteria or viruses can also be genetically modified and transformed into specific nutrient delivery vectors to target specific regions of the gastrointestinal tract, thus allowing more efficient absorption of nutrients before reaching the brain. This approach offers considerable potential for more precise nutrient delivery, thereby improving the effectiveness of nutritional interventions while opening up new avenues for the prevention of neurological diseases.

Personalized Diets

Personalized diets leverage advances in biotechnology to understand the specific nutritional needs of each individual.

One of the key advances in this context lies in the analysis of the individual genome. Thus, nutritional genomics uses genetic information to identify specific factors related to brain health, such as the response to nutrients such as omega-3 fatty acids, antioxidants, or B vitamins. This information is then used to design personalized dietary plans that meet the specific nutritional needs of each individual.

Furthermore, precise analysis of the microbiota composition, made possible by genomic sequencing, allows for the quantification of the different bacterial species present in the gut specific to each individual. Thus, healthcare professionals can obtain valuable information about digestive health, immunity, and even metabolism. This paves the way for more targeted interventions, such as specific dietary plans, personalized probiotics, or other therapies aimed at restoring and maintaining optimal microbiota balance on an individual basis.

❖ Future Perspectives

Future prospects for nutrition in the context of brain health focus on identifying nutrients that promote neurogenesis and prevent neurodegenerative diseases. Integrative medical approaches emphasize the importance of nutrition for both physical and mental health, with personalized dietary prescriptions becoming essential.

Current Research on Nutrition and Brain Health

Scientists are intensifying their research to understand the complex interactions between specific food components and brain processes. Studies focus on identifying foods that promote neurogenesis, synaptic plasticity, and the prevention of neurodegenerative diseases. Technological advances, such as advanced brain imaging and molecular analysis tools, facilitate a deeper understanding of the effects of diet on brain structure and function. These methods allow for a more precise assessment of individual responses to specific dietary regimes.

Integrative Medical Approaches

Diet has become a cornerstone of integrative medical approaches. Healthcare professionals recognize the importance of nutrition not only for physical health but also for mental health. Thus, personalized dietary prescriptions become an essential component of treatment plans. The emergence of integrative care programs often involves collaboration among nutritionists, neuroscientists, psychologists, and other healthcare professionals.

Long-Term Implications of Personalized Nutrition

Personalized nutrition can play a crucial role in preventing neurological disorders. By adapting diet according to genetic predispositions and individual biomarkers, it is possible to reduce the risks of diseases such as Alzheimer's and Parkinson's.

Beyond disease prevention, personalized nutrition also offers perspectives for enhancing cognitive performance. Specific dietary regimes can be designed to optimize concentration, memory, and other aspects of brain function, thus contributing to overall better mental health.

Biotechnology for Brain Longevity

This chapter explores current and future developments in biotechnology aimed at extending brain longevity, highlighting technical advancements, ethical considerations, and perspectives on how this could reshape our understanding of brain life.

❖ Biotechnology for Brain Longevity

Throughout the 20th century, biotechnology has played a central role in research on brain longevity.

Historical Foundations

The mid-20th century witnessed major discoveries that shaped biotechnology. Among these advancements, the discovery of DNA and the revolution in molecular biology opened uncharted doors. Understanding genetics allowed scientists to explore the molecular foundations of aging.

In the 1980s and 1990s, technological progress such as complete sequencing of the human genome also revolutionized research. These tools provided a detailed map of genes related to aging and longevity, enabling scientists to specifically target biological processes involved in age-related brain deterioration.

Discoveries also highlighted the crucial role of telomeres, the protective ends of chromosomes, in cellular aging. Advanced technologies emerged to manipulate these structures, thus opening new pathways to delay the process of brain aging.

Technological Advances and Integration of Neuroscience

The last decade has been characterized by an increasing integration of neuroscience, genomics, and biotechnology. Advanced brain imaging techniques, such as fMRI, allow researchers to track brain changes related to longevity in real-time. Additionally, artificial intelligence-based approaches are used to analyze vast datasets and identify complex correlations.

These technological advances have also led to innovative clinical trials, exploring early interventions to slow down brain aging. Drugs and gene therapies specifically designed to target aging mechanisms are in development, opening promising prospects for the future.

❖ Brain Preservation and Regeneration

Techniques for brain preservation and regeneration represent a fascinating field of medical research, aiming to develop innovative approaches to protect and repair brain tissue. These promising methods pave the way for significant advances in the treatment of neurological diseases and brain injuries.

Neuronal Preservation

Neuronal preservation aims to maintain the health and function of brain cells, thereby protecting brain tissue against stress, damage, and premature aging. Several methods have been developed to achieve this goal.

Antioxidant Therapies

Free radicals, natural byproducts of metabolism, can damage brain cells. Antioxidant therapies, such as the administration of vitamins C and E, seek to neutralize these free radicals, thereby reducing oxidative stress and preserving neuronal health. Studies have shown that these therapies may contribute to the prevention of age-related neurodegenerative diseases.

Cryopreservation Techniques

Cryopreservation involves preserving brain tissues at extremely low temperatures. This technique is often used in the field of medical research and neurobiology to store brain samples, but it also explores potential applications in organ preservation for transplantation. However, challenges associated with cryopreserving the entire human brain for medical applications are complex, including potential cellular damage during the thawing process.

Metabolic Modulation

Metabolic modulation aims to influence metabolic processes at the cellular level to optimize neuronal health. This may include changes in diet, physical exercise, and other strategies aimed at improving the energy management of brain cells. Studies have suggested that specific diets, such as the Mediterranean diet, may have beneficial effects on neuronal preservation.

Cellular Regeneration

Cellular regeneration, unlike preservation, aims to repair and restore damaged or lost brain cells. This innovative approach offers considerable potential for treating neurodegenerative diseases and brain injuries.

Gene Therapies

Gene therapies seek to introduce specific genes into the brain to stimulate cell growth and regeneration. Studies have explored the use of viral vectors to deliver cell growth-promoting genes into specific areas of the brain. However, concerns remain regarding the safety and effectiveness of this approach.

Neurostimulation

Neurostimulation involves using electrical or magnetic stimuli to influence neuronal activity and promote regeneration. Techniques such as transcranial direct current stimulation (tDCS) and transcranial magnetic stimulation (TMS) are being studied for their ability to modulate brain plasticity and promote cellular regeneration.

Growth Factors

Growth factors are proteins that play a crucial role in the development, differentiation, and survival of cells. By administering growth factors, such as brain-derived neurotrophic factor (BDNF), researchers aim to promote these processes in the brain, which could have beneficial implications for neuronal health, particularly in the context of tissue regeneration and protection.

Cellular Therapies

Cellular therapies, including stem cell transplantation, aim to harness the intrinsic capacity of regenerative cells to replace those that are lost or damaged in the brain. In this context, stem cells, which are undifferentiated and versatile cells, are implanted into the brain with the hope that they will differentiate into specific cells needed for regeneration or repair of brain tissues. This approach explores the natural regeneration potential of cells to treat neurological conditions or brain injuries.

Challenges Associated with Brain Preservation and Regeneration

Brain preservation and regeneration present unique challenges, including coordinating complex biological processes and minimizing potential risks:

- *Temporal and Spatial Coordination:* Harmonizing preservation and regeneration processes to meet the temporal and spatial needs of the brain is a major challenge. Precise synchronization of interventions is crucial to optimize outcomes and minimize potential risks.

- *Potential Risks of Cellular Regeneration:* Although cellular regeneration offers considerable potential, concerns remain regarding potential risks, such as tumor formation or deregulation of cell growth. Strict protocols and thorough safety assessments are necessary to ensure that cellular regeneration approaches do not have undesirable consequences.

- *Complexity of Cellular Interactions:* Understanding the complexity of cellular interactions in the brain is a constant challenge. Signaling mechanisms and cellular responses can vary significantly depending on specific conditions, making personalized approaches essential to maximize effectiveness.

❖ Genetic Factors and Brain Longevity

Understanding the genetic factors of brain longevity involves identifying the exact genes that play a crucial role in maintaining neuronal health throughout life.

Identification of Genes Associated with Brain Longevity

Genetic Association Studies

Genetic association studies analyze the DNA of entire populations to identify genetic variations associated with increased brain longevity. These studies have led to the discovery of several genes that seem to play a key role in the protection and preservation of brain cells. For example, the APOE gene has been associated with a greater susceptibility to neurodegenerative diseases such as Alzheimer's disease.

Whole Genome Sequencing and Omics Approaches

Whole genome sequencing has opened up new possibilities by allowing comprehensive mapping. Omics approaches, such as genomics, transcriptomics, and proteomics, also offer a deeper understanding of underlying molecular processes. These techniques identify specific genetic biomarkers associated with exceptional brain longevity.

Epigenetics and Brain Longevity

Epigenetics, the study of chemical modifications that regulate gene expression without altering their sequence, also plays a crucial role in brain longevity. Studies have revealed how epigenetic changes can influence gene regulation related to neuronal health and longevity.

Genetic Manipulation to Promote Neuronal Longevity

Genetic knowledge allows for the consideration of genetic manipulation approaches to promote neuronal longevity by stimulating or genetically modifying mechanisms that preserve and protect brain cells.

Gene Therapies for Neuronal Growth

Gene therapies aim to introduce specific genes promoting neuronal growth into the brain. Experiments have shown that introducing genes responsible for the production of growth factors could stimulate cellular regeneration and the formation of new synaptic connections.

Modulation of Cellular Senescence Processes

Cellular senescence, a state in which cells cease to divide and enter a phase of accelerated aging, is closely linked to brain aging. Approaches aim to genetically modulate these processes to extend the functional lifespan of brain cells.

Correction of Genetic Mutations

Genetic editing, especially with technologies like CRISPR-Cas9, opens the possibility of correcting specific genetic mutations associated with neurodegenerative diseases and accelerated aging processes. This targeted approach offers considerable potential for promoting neuronal longevity.

❖ **Prevention of Neurodegenerative Diseases**

Prevention of neurodegenerative diseases relies on significant advances in understanding the molecular, genetic, and cellular mechanisms involved in the development of these conditions. Several approaches have been developed to target these mechanisms and prevent disease progression.

Utilization of Biotechnology to Prevent Neurodegenerative Diseases

Gene Therapies and Gene Expression Modulation

Gene therapies aim to introduce specific genes or modulate gene expression to prevent detrimental changes leading to neurodegenerative diseases. Studies have examined the possibility of regulating the expression of certain genes involved in neuroprotection, thus offering a potentially preventive approach.

Inhibition of Protein Aggregates and Inflammatory Processes

Neurodegenerative diseases are often characterized by the accumulation of protein aggregates and inflammatory processes in the brain. Biotechnology explores approaches to inhibit the formation of these aggregates and modulate the inflammatory response, offering prevention possibilities.

Early Diagnosis through Biotechnology

Biotechnology also plays a crucial role in the early diagnosis of neurodegenerative diseases, allowing early intervention before severe symptoms manifest.

Biomarkers and Advanced Brain Imaging

The search for specific biomarkers in blood, cerebrospinal fluid, or even advanced brain imaging, such as fMRI, allows the detection of subtle changes in the brain before the onset of obvious symptoms. These biomarkers provide valuable predictive clues for neurodegenerative diseases.

Omics Analysis for Early Genetic Signature

Omics approaches, such as genomics and proteomics, are used to analyze the early genetic signature associated with neurodegenerative diseases. These techniques provide a deep understanding of early molecular alterations that can be used as predictive indicators.

Artificial Intelligence and Early Diagnosis

Artificial intelligence (AI) is increasingly used in the analysis of complex data related to neurodegenerative diseases. AI algorithms can detect subtle patterns in omics and imaging datasets, allowing early diagnosis and targeted intervention.

Implications of Preventing Neurodegenerative Diseases on Brain Longevity

Preventing neurodegenerative diseases through biotechnology has profound implications for brain longevity, defined as the functional and healthy lifespan of the brain.

Improvement of Brain Quality of Life

Successful prevention of neurodegenerative diseases contributes to improving brain quality of life by preserving cognitive functions and avoiding severe neurological declines. This has positive implications for maintaining independence and actively participating in society.

Reduction of Healthcare Costs

Preventing neurodegenerative diseases can also lead to a significant reduction in healthcare costs. Preventive treatments can avoid the need for long-term care and intensive treatments, thereby relieving the financial burden on healthcare systems.

❖ Ethical Challenges of Brain Longevity

Ethical Dilemmas Related to Prolonging Brain Life

Prolonging brain life raises complex ethical dilemmas that question our understanding of life, death, and individual autonomy.

Definition of Brain Life and Death

Prolonging brain life challenges the traditional definition of death, usually associated with the cessation of cardiac or respiratory functions. When the brain can be artificially kept alive, it becomes necessary to redefine the criteria for brain life and death, raising questions about when it is legitimately declared that a person has passed away.

Autonomy and Decision Making

The issue of individual autonomy is central in the context of brain longevity. Individuals must be able to make informed decisions regarding the prolongation of their brain life, raising delicate questions about mental capacity, individual preferences, and the role of loved ones in the decision-making process.

Public Education

Informed consent requires a thorough understanding of the benefits, risks, and ethical implications of interventions aimed at prolonging brain life. Public education becomes a crucial element to ensure that individuals are adequately informed to make informed decisions, thereby avoiding potential manipulation or decision-making based on incomplete information.

Complex Medical Scenarios

In complex medical scenarios, such as severe accidents or neurodegenerative diseases, decision-making often becomes a shared responsibility among the patient, loved ones, and healthcare professionals. Advance directives and pre-discussions can help clarify the patient's wishes, but ethical dilemmas remain regarding how to make decisions in the patient's best interest.

Social Implications

Economic Disparities and Access to Longevity Technologies

Brain longevity technologies may become inaccessible to certain populations due to economic factors. This raises questions about social justice and health equity, as only economically privileged individuals may have access to these interventions.

Genetic Inequalities and Health Implications

The emergence of gene therapies and genetic editing for brain longevity raises concerns about creating genetic disparities between individuals who have access to these technologies and those who do not.

Impact on Social and Economic Structure

Prolonging brain life can potentially reshape social and economic structure. Populations living longer could have implications for the labor market, social security systems, and other aspects of social life, creating complex challenges for society.

❖ Future Perspectives

Rapid advancements in neuroscience and biotechnology hold the promise of major breakthroughs that could redefine how we perceive and manage brain longevity.

- *Personalized Gene Therapies:* Future breakthroughs could see the emergence of personalized gene therapies, specifically targeting genetic factors related to brain longevity. The ability to selectively modify genes responsible for brain health could open up unprecedented opportunities for the prevention and treatment of neurodegenerative diseases.

- *Nanotechnology for Brain Repair:* Nanotechnology could play a crucial role in repairing brain injuries and neuronal regeneration. Nanorobots could be designed to target damaged areas of the brain, repairing neuronal connections and thereby improving brain function.

- *Advanced Brain-Machine Interfaces:* Brain-machine interfaces are expected to undergo significant improvements, allowing for smoother communication between the brain and external devices. These interfaces could facilitate the restoration of lost functions, improving the quality of life for people with brain impairments.

- *Precise and Personalized Neurostimulation:* Advancements in neurostimulation could lead to more precise and personalized techniques. Deep brain stimulation and other forms of neurostimulation could be more specifically tailored to individual needs, thereby improving the effectiveness of treatments.

3

Optimizing the Brain: Strategies to Enhance Cognition

Neuroplasticity for Brain Remodeling

This chapter explores the mechanisms of neuroplasticity, innovative techniques to stimulate it, and the implications for achieving exceptional human performance through optimized brain plasticity.

❖ Understanding Neuroplasticity

Understanding and harnessing neuroplasticity, the remarkable ability of the brain to adapt and reorganize, opens fascinating prospects in the field of neuroscience.

Foundations of Neuroplasticity

Brain Plasticity

Neuroplasticity, or brain plasticity, is a key concept in the field of neuroscience. It is defined as the brain's capacity to adapt and reorganize in response to experience, environmental stimuli, and injuries. This remarkable property allows neuronal connections to strengthen, form, or modify, thus influencing brain structure and function.

Neuroplasticity can be observed at different levels, ranging from molecular and cellular changes to large-scale adjustments in neuronal networks. Understanding these mechanisms offers important insights into the treatment of neurological disorders, the development of rehabilitation strategies, and the promotion of mental well-being.

Study Techniques

The study of neuroplasticity relies on various techniques. Among them, functional brain imaging, such as fMRI, provides real-time visualization of brain regions activated during specific tasks, enabling the identification of functional changes related to neuroplasticity.

Electrophysiological methods, such as EEG and magnetoencephalography (MEG), record the brain's electrical activity, allowing observation of changes in brain wave patterns associated with plasticity.

Studies on patients with brain injuries, combined with non-invasive brain stimulation techniques such as transcranial magnetic stimulation or deep brain stimulation, offer insights into plasticity related to functional recovery.

Furthermore, molecular and cellular methods, such as molecular biology and optogenetics, allow exploration of the underlying mechanisms of plasticity at the level of synapses and neuronal circuits.

Brain Adaptability at Different Life Stages

Once considered limited to early childhood, neuroplasticity is now understood as a dynamic process that persists throughout life:

- Childhood is marked by intense neuroplasticity, allowing the brain to quickly adapt to a constantly changing environment. It is a crucial period for learning fundamental skills and developing cognitive foundations.
- Although neuroplasticity slightly decreases with age, it persists into adulthood. Studies show that adults can

continue to learn new skills, adapt their behavior, and modify their thinking patterns through neuroplasticity.

- Even during aging, the brain retains some capacity for plasticity. Strategies such as physical exercise, lifelong learning, and cognitive stimulation can promote neuroplasticity in older adults, thus contributing to the preservation of brain functions.

Biological Mechanisms

Synaptic Strengthening and Pruning

The dynamic combination of synaptic strengthening and pruning is essential for optimizing the efficiency of the neuronal network.

- Synaptic Strengthening, resulting from regular activation of a synapse, amplifies neuronal transmission by promoting increased neurotransmitter release and strengthening of the connection. This phenomenon occurs when neuronal activity is repeated and persistent. Thus, when one neuron regularly sends signals to another, the strength of the synaptic connection between them increases.

- Concurrently, synaptic pruning occurs by eliminating less used or less functional synaptic connections. During neuronal development, an excess of synapses is initially formed, creating a sort of "synaptic overpopulation." Synaptic pruning then occurs, selectively eliminating some synaptic connections. This phenomenon helps shape the structure and efficiency of neuronal networks, promoting the consolidation of the most relevant synaptic pathways and thus contributing to the optimization of neuronal circuits. Synaptic pruning is particularly significant during critical periods of brain development, but it also persists throughout life, allowing the brain to adapt to environmental changes, optimize its performance, and promote continuous learning.

Neurogenesis

Neurogenesis refers to the process of forming new neurons, a capacity that was long overlooked but is now established as an essential feature of the brain. Thus, contrary to the widespread belief that the number of neurons was set at birth, recent studies have demonstrated that neurogenesis also occurs in adults, mainly in two regions of the brain: the hippocampus, crucial for memory and learning, and the olfactory gyrus, associated with olfactory perception.

This complex process involves the proliferation, migration, differentiation, and functional integration of new neurons into pre-existing networks. Factors such as physical exercise, environmental enrichment, and even sleep appear to positively influence neurogenesis. Furthermore, neurogenesis has significant implications for brain plasticity, mood regulation, and the ability to adapt to new cognitive challenges.

These findings have promising implications for the treatment of neurological and psychiatric disorders, offering a new perspective on adult brain plasticity and its potential for regeneration.

Structural Remodeling

Studies have shown that stimulating experiences, whether intellectual, social, or physical, can lead to morphological and functional changes in the brain.

The hippocampus, for example, a key region involved in memory and learning, is particularly sensitive to these influences. When an individual is exposed to cognitive challenges or enriching environments, the number of synaptic connections may increase, thereby promoting neuronal plasticity. Additionally, neurotrophins, growth factors essential for the survival and development of neurons, are often released in response to these stimuli, facilitating the formation of new connections.

This structural plasticity is not limited to a critical period of development but persists throughout life. Activities such as learning new skills, regular physical activity, or even exposure to stimulating social environments can induce beneficial changes in brain structure.

Thus, structural remodeling offers considerable hope for interventions aimed at improving cognitive abilities, treating neurological disorders, and promoting mental well-being.

Practical Applications

Rehabilitation After Brain Injury

Understanding neuroplasticity has significant implications in rehabilitation after a brain injury. Rehabilitation programs harness brain plasticity to promote functional recovery by encouraging the reorganization of damaged neuronal circuits.

Learning and Education

In the field of education, considering neuroplasticity offers more effective teaching approaches. Teaching methods that actively engage the brain, foster emotional involvement, and encourage regular practice capitalize on brain plasticity to optimize learning.

Treatment of Neurological and Psychiatric Disorders

Stimulating neuroplasticity opens promising avenues for the treatment of neurological and psychiatric disorders such as depression and Parkinson's disease. Approaches like cognitive-behavioral therapy exploit brain plasticity to reorganize thought patterns and improve symptoms.

❖ Enhancement of Neuroplasticity

Certain techniques optimize brain performance by stimulating brain adaptability, thus promoting significant improvements in cognition and neuronal plasticity. These methods are promising for optimizing learning and memory, as well as in the field of medical neurological rehabilitation.

Traditional Methods

Experiential Learning

Experiential learning remains one of the most fundamental methods for stimulating neuroplasticity. By actively engaging in various activities, new skills are acquired, triggering structural and functional changes in the brain.

Repetition and Deliberate Practice

Constant repetition and deliberate practice are key elements in forming new synaptic connections. These traditional methods reinforce neuronal pathways, facilitating the automation of skills and knowledge.

Physical Exercise

Regular physical exercise has positive effects on neuroplasticity, stimulating the growth of new neurons and promoting synaptic plasticity.

Sensory Stimulation

Sensory stimulation, whether visual, auditory, or tactile, enriches the sensory environment, encouraging the diversity of neuronal connections.

Technological Innovations

Transcranial Magnetic Stimulation

Transcranial Magnetic Stimulation (TMS) is an innovative neuromodulation technique based on the fundamental principle of electromagnetic induction.

Thus, electromagnetic coils placed on the scalp generate variable magnetic fields, creating electrical currents in targeted brain regions. These currents alter neuronal activity, triggering short-term or long-term effects.

Different brain regions can be targeted by adjusting the position and orientation of the coils. Researchers can thus target specific areas related to particular cognitive functions or regions associated with neurological or psychiatric disorders.

Here are some examples of TMS applications:

- Depression Treatment: TMS has received considerable attention for its potential use in treating treatment-resistant depression. Stimulation of the dorsolateral prefrontal cortex, involved in emotional regulation, has shown promising results in improving mood in some patients.

- Chronic Pain Relief: Studies have explored the effectiveness of TMS in relieving chronic pain, including migraines and fibromyalgia. By modulating the neural circuits involved in pain perception, TMS offers a potentially non-pharmacological alternative for pain management.

- Stroke Rehabilitation: TMS is also being studied in the context of rehabilitation after a stroke. By targeting brain regions associated with movement, stimulation can facilitate brain plasticity and improve motor recovery in individuals who have had a stroke.

- Cognitive Function Modulation: Studies have examined how stimulation of certain brain regions can influence memory, attention, and other cognitive processes, paving the way for innovative approaches to improving mental performance.

Neurofeedback

Neurofeedback, also known as neurotherapy or biofeedback, is emerging as an innovative approach to stimulate brain plasticity.

This technique involves modulating brain waves to optimize each individual's potential and relies on real-time feedback of brain activity. Thus, sensors record brain waves, such as alpha, beta, delta, and theta waves, and provide instant feedback to the individual about their mental state. Participants learn to modify their own brain wave patterns by receiving positive reinforcement when they reach specific states. This promotes self-regulation by encouraging the brain to produce wave patterns associated with concentration, relaxation, or other defined goals. They can be traditional or automated by applications.

Here are some examples of neurofeedback applications:

- Cognitive Performance Enhancement: Neurofeedback is used to improve various aspects of cognitive performance. Targeted sessions can promote increased attention, memory, and problem-solving by modulating brain plasticity associated with these functions.

- Stress and Anxiety Management: Modulating brain waves via neurofeedback can be beneficial for managing stress and anxiety. By promoting brain states associated with relaxation, this technique can help reduce stress levels and improve emotional well-being.

- Treatment of Neurological and Psychiatric Disorders: Neurofeedback is explored as a complementary treatment option for disorders such as attention deficit hyperactivity disorder (ADHD), autism spectrum disorders, and mood disorders. By modifying brain wave patterns, it aims to alleviate associated symptoms.

Here are some techniques and protocols of neurofeedback:

- Sensorimotor Neurofeedback: This approach focuses on regulating brain waves in relation to body movements. It can be used in stroke rehabilitation or to improve coordination and physical performance.

- Functional Magnetic Resonance Imaging (fMRI) Neurofeedback: The use of fMRI allows the identification of brain regions active during neurofeedback, providing a more precise understanding of neuroplastic changes induced by this technique.
- Visual System-Based Neurofeedback: By using visual stimuli, visual system-based neurofeedback aims to improve the regulation of brain waves associated with visual perception, offering potential benefits for vision and perception.

Although the sustainability of neurofeedback effects still requires further investigation, preliminary results suggest that benefits may persist beyond training sessions and induce long-term changes in brain structure and function. This includes adaptations at the level of neuronal connections and modifications in brain regions involved in targeted functions.

Mobile Applications

The use of mobile applications is emerging as a contemporary approach to stimulate brain plasticity. These applications, often based on interactive games, offer opportunities for personalized training, allowing users to engage in activities designed to promote brain flexibility and adaptability in a playful manner. Virtual reality headsets can also be used to create immersive experiences that engage various cognitive functions. Indeed, virtual environments can be designed to provide complex visual, auditory, and kinesthetic stimuli, engaging multiple sensory modalities to maximize brain changes. However, it is crucial to assess the scientific validity of these applications to ensure their actual effectiveness.

Other Approaches

Combining Physical Exercise and Cognitive Stimulation

The combination of physical exercise with complex cognitive tasks creates a powerful synergy. Studies suggest that physical exercise increases the release of neurotrophic factors, promoting neuron growth and survival, while cognitive tasks strengthen synaptic connections.

Relaxation and Meditation Techniques

Relaxation and meditation techniques, although traditional, have been recently validated by modern research for their impact on neuroplasticity. These practices promote emotional regulation and concentration, beneficially altering brain activation patterns.

❖ Exceptional Human Performance

Neuroplasticity offers the prospect of achieving exceptional human performance by enabling the brain to adapt and develop in response to specific experiences and training.

Brain Plasticity and Excellence

Brain Plasticity and Music

Brain plasticity and music are closely linked, offering a fascinating exploration of the adaptable capabilities of the human brain. In-depth studies have revealed that intensive musical practice can induce structural and functional changes in the brain, demonstrating remarkable neuronal plasticity. Musicians, in particular, show anatomical modifications in brain regions related to fine motor skills, auditory memory, and sensorimotor coordination. For example, the primary motor cortex and the corpus callosum, which facilitates communication between the brain hemispheres, may show expansion in seasoned musicians. Moreover, music-related brain plasticity is not limited to professionals; even amateurs engaging in regular musical learning can benefit from beneficial changes.

Music, as a complex stimulus, engages various brain regions, prompting the brain to reorganize to process this information optimally. This plasticity can have significant implications, ranging from cognitive skill enhancement to functional recovery after brain injuries. Rehabilitation programs using music are increasingly used to help individuals with neurological disorders regain motor and cognitive abilities. Thus, studying brain plasticity in the musical context offers a captivating perspective on how engagement with sound art can shape and reshape the very structure of our brain, opening doors to new avenues for therapy and brain performance enhancement.

Brain Plasticity and Eidetic Memory

Brain plasticity and eidetic memory, also known as photographic memory, form a captivating research field that explores the brain's ability to adapt and retain information exceptionally. Eidetic memory is characterized by the ability to remember visual details with remarkable accuracy after brief exposure. Studies suggest that this ability is influenced by brain plasticity, as it involves changes in neuronal connections and brain regions responsible for visual processing and short-term memory.

Individuals with eidetic memory seem to exhibit adaptations in brain areas related to visual perception, such as the occipital cortex and temporal cortex, where visual memory is primarily processed. Regular practice of visual memory reinforcement techniques can stimulate this plasticity, thereby enhancing the ability to retain images with great precision. However, it is worth noting that eidetic memory is rare, and its existence in some individuals underscores the diversity of human memory abilities.

Understanding the relationship between brain plasticity and eidetic memory opens intriguing perspectives for the development of educational strategies and memorization techniques. By exploring how the brain reacts and adapts to exceptional memory abilities, research on brain plasticity and eidetic memory offers fascinating insights to enhance our understanding of the complex functioning of the organ that governs our ability to perceive and remember the world around us.

Brain Plasticity and Athletic Performance

Brain plasticity and athletic performance form a dynamic partnership in which the brain adapts to improve athletes' coordination, concentration, and motor skills. Studies have shown that intensive physical training can induce structural changes in the brain, particularly in areas associated with motor control, movement planning, and sensory perception. These neuronal adaptations enable athletes to develop faster reflexes, better coordination, and enhanced learning ability for specific movements.

Brain plasticity also offers benefits in the realm of recovery after sports injuries. Athletes who suffer injuries can use neurological rehabilitation to strengthen brain connections associated with movement, thereby speeding up the healing process and minimizing the loss of athletic skills.

The cognitive dimension of brain plasticity is also crucial for athletic performance. Concentration, quick decision-making, and stress management are all mental skills that can be improved through specific mental training. Techniques such as mental visualization, meditation, and other forms of mental preparation can stimulate brain plasticity, promoting optimal athletic performance.

In summary, brain plasticity and athletic performance have a bidirectional relationship, with each domain significantly influencing the other. Understanding how the brain adapts and changes in response to physical training opens doors to maximize athletes' potential and explore new frontiers in enhancing sports performance.

Examples of Notable Brain Plasticity

Exceptional cases of brain plasticity shed light on situations where the human brain has demonstrated remarkable adaptability, often in the face of severe medical conditions. Here are some notable examples:

1. Phantom Limb Pain and Mirror Box Therapy: Amputees may experience intense pain in the body part that is no longer there, known as phantom limb pain. Mirror box therapy, a technique based on brain plasticity, has shown promising results in helping relieve this pain by persuading the brain that it still controls the missing limb.

2. Rehabilitation after Traumatic Brain Injury: Individuals who have suffered traumatic brain injury can benefit from brain plasticity to regain certain cognitive functions. Specific rehabilitation programs can stimulate neuronal reorganization.
3. Recovery after Hemispherectomy: Some individuals, often children, have undergone hemispherectomy, a surgery where one half of the brain is removed to treat disorders such as severe epilepsy. Despite the loss of a substantial part of the brain, some patients have shown remarkable ability to recover functions such as motor skills and language.
4. Adaptation to Blindness: Blind individuals have demonstrated astonishing brain plasticity by using areas of the brain that are normally dedicated to vision to process other sensory information, such as touch and hearing, thereby compensating for the loss of sight.
5. Children recovering from damaged frontal lobes: In children, the brain has particularly high plasticity. Some children who have sustained injuries to the frontal lobe have shown notable recovery of cognitive functions, often through neuronal reorganization.
6. Chronic Pain Treatment: Studies have suggested that meditation and other non-pharmacological approaches can induce changes in the brain that alter pain perception, illustrating brain plasticity in chronic pain management.

Here are some examples illustrating brain plasticity in individuals:

1. Phineas Gage: Although his case dates back to the 19th century, Phineas Gage is often cited in discussions about brain plasticity. After a workplace accident in which an iron rod pierced his brain, Gage survived, but his behavior changed. This was interpreted as an early illustration of the brain's ability to adapt to severe injuries.

2. Gabby Giffords: The former member of the U.S. Congress survived an assassination attempt in 2011, during which she was severely injured in the head. She has since demonstrated incredible brain plasticity by recovering motor and cognitive functions.

3. Ian Waterman: After a viral infection damaged his nerves, Ian Waterman lost sensation and control over much of his body. However, through remarkable adaptation, he learned to control his movements primarily based on vision.

4. Matt Wetschler: After a freediving accident resulting in a spinal cord injury, Matt Wetschler used brain plasticity to regain some motor and sensory functions. He shared his journey in the documentary "Diving Into the Unknown."

5. Michelle Mack: After undergoing hemispherectomy at the age of six to treat epilepsy, Michelle Mack showed remarkable ability to recover functions and lead a relatively normal life.

6. Edwyn Collins: Musician Edwyn Collins suffered two strokes that affected his ability to speak and play music. Through intensive rehabilitation and neuroplasticity, he was able to regain some of his musical skills.

❖ Limits and Precautions

Challenges and Future Perspectives of Neuroplasticity

Limits of Neuroplasticity

While neuroplasticity offers remarkable opportunities, it has its limits. Some patterns of thinking and behaviors are more difficult to change due to the long-term stabilization of neuronal connections. Understanding these limits is essential for designing effective interventions.

Disruption of Brain Balance

Excessive manipulation of neuroplasticity could lead to imbalances in brain function. Overly aggressive interventions could disrupt natural neural circuits, potentially leading to cognitive, emotional, or even psychiatric disorders.

Unexpected Side Effects

Methods aimed at stimulating neuroplasticity could have unexpected side effects. Changes in one brain region could have repercussions on others, creating unforeseen consequences that require careful attention.

Ethical Debates

The possibility of intentionally modulating neuroplasticity raises important ethical questions. Interventions aimed at enhancing cognition or treating mental disorders through the modulation of brain plasticity require rigorous evaluation of risks and benefits.

Equity in Access and Use

Ethical debates around neuroplasticity include concerns about equity in access and use. If some methods are reserved for an elite, it could exacerbate inequalities and create a disparity in access to the benefits of cognitive enhancement.

Informed Consent and Autonomy

Interventions aimed at modifying neuroplasticity raise crucial questions of informed consent and autonomy. Individuals must be fully informed of potential risks and benefits, and their consent must be freely given without external pressure.

Ethical Boundaries of Performance Enhancement

Research and the use of extreme methods to enhance cognitive performance question the boundaries of ethics. How far can one go to increase human capabilities before crossing moral and ethical lines?

Individual Freedom and Responsibility

The debate between individual freedom and collective responsibility is central in the exploration of neuroplasticity. Do individuals have the right to modify their brains according to their personal desires, or does society have the responsibility to regulate these modifications to avoid unforeseen consequences and negative social impacts?

Ethical and Legislative Oversight

Strong ethical and legislative oversight is necessary to guide the use of methods aimed at modifying neuroplasticity. Clear ethical standards and legislative regulations can help protect individuals while allowing responsible exploration.

Education and Awareness

Education and awareness are key elements in informing the public about the implications of modifying neuroplasticity. A high level of understanding contributes to informed consent and fosters informed discussions on associated ethical issues.

❖ Future Perspectives

Technological Advances in Brain Imaging

Continuous advancements in brain imaging technologies, such as high-resolution functional MRI and real-time neuroimaging, allow for finer exploration of the mechanisms of neuroplasticity. This opens the door to a deeper understanding of brain changes in response to specific stimuli.

Emergence of Targeted Therapies

Current research is focusing on the development of targeted therapies harnessing neuroplasticity to treat specific neurological disorders. Innovative approaches aim to stimulate specific brain regions to restore impaired functions, offering new perspectives for rehabilitation after brain injuries.

Genomics and Neuroplasticity

Genomics is coming into play to elucidate the genetic aspects of neuroplasticity. Understanding genetic variations influencing brain plasticity opens up possibilities to personalize interventions based on each individual's genetic profile.

Enhancing Cognition through Genetic Editing

This chapter explores the possibilities and ethical dilemmas associated with brain modification through genetic editing to improve cognitive abilities.

❖ Genetic Editing and Brain Modification

Foundations of Genetic Editing

Genetic editing is a technique that allows for precisely modifying the genetic material of an organism by introducing targeted changes into its DNA. This modification can involve the insertion, deletion, or substitution of specific genetic sequences.

One of the most widely used genetic editing technologies is CRISPR-Cas9, which utilizes a protein (Cas9) to cut DNA at precise locations, allowing researchers to introduce desired genetic modifications.

Genetic editing has diverse applications, from fundamental research in biology to modifying plants to enhance disease resistance, as well as in the medical field to treat specific genetic diseases. This technology offers exciting prospects while raising significant ethical questions, particularly those related to manipulating the human genome and its long-term implications.

In brain research, genetic editing enables the specific study of the role of certain genes in development, function, and neurological disorders. Experiments on genetically modified animal models can help elucidate the links between genes and complex traits such as memory, learning, and even neurodegenerative conditions.

Genetic Editing Technologies

CRISPR-Cas9

CRISPR-Cas9 technology has revolutionized genetic editing by offering exceptional precision and versatility. By using a protein guided by RNA, CRISPR-Cas9 can specifically target DNA sequences, facilitating gene modification with increased efficiency.

CRISPR-Cas9, an acronym for "Clustered Regularly Interspaced Short Palindromic Repeats" and "CRISPR-associated protein 9", is a revolutionary genetic editing method that allows for the specific modification of an organism's genome. This technique utilizes a protein called Cas9 as "molecular scissors" to cut DNA at precise locations, determined by complementary RNA guide sequences to targeted genomic sequences. Once the DNA is cut, the organism's cells naturally repair the break, but this can result in the insertion, deletion, or substitution of specific genetic sequences.

CRISPR-Cas9 has transformed the field of genetic editing by providing a faster, less costly, and more precise method than previous techniques. This technology is widely used in biomedical research to study genetic functions, develop disease models, and potentially treat genetic diseases. It also has applications in agriculture, enabling the creation of disease-resistant plants, and in other areas such as synthetic biology.

Although CRISPR-Cas9 offers exciting opportunities, it also raises ethical questions and concerns about its potential use to modify the human genome in non-therapeutic ways. Due to its power and potential, its use requires strict ethical standards and careful considerations regarding its impact on society.

Other Genetic Editing Tools

In addition to CRISPR-Cas9, other genetic editing tools, such as TALENs (Transcription Activator-Like Effector Nucleases) and meganucleases, also present specific advantages. Each of these tools has its own characteristics, allowing adaptability to different situations and goals.

TALENs and meganucleases are also genetic editing tools that, like CRISPR-Cas9, allow for precise modification of an organism's genome. Each of these systems has specific characteristics.

1. **TALENs:** TALENs are genetic editing enzymes based on transcription activator-like effector (TALE) proteins. These proteins are derived from pathogenic bacteria that infect plants. To create a TALEN, specific TALE sequences are fused with restriction enzymes, forming a molecule capable of cutting DNA at precise locations. TALENs can be designed to target specific genomic sequences, making them similar to CRISPR-Cas9 in their ability to target specific genome regions.

2. **Meganucleases:** Meganucleases are genetic editing enzymes derived from bacterial proteins. They have the ability to recognize and cut specific DNA sequences. Meganucleases are known for their high specificity, meaning they are less likely to cause

unintended cuts in the genome. However, they may be more challenging to design and use than other genetic editing technologies.

While CRISPR-Cas9 is currently the most widespread genetic editing method due to its simplicity and efficiency, TALENs and meganucleases have been widely used in research and continue to be potentially valuable options. Each of these technologies has its own advantages and limitations, and the choice between them often depends on the specific context of the experiment or application.

Limitations of Current Technologies

Despite their power, genetic editing technologies present challenges, including issues of efficiency and specificity: unintended modifications. These challenges underscore the need for ongoing research to improve the accuracy and safety of these tools.

Despite the significant advances it represents, genetic editing technology, especially CRISPR-Cas9, has certain limitations and technical challenges. Here are some of the main current limitations:

- Precision: Although CRISPR-Cas9 technology is highly precise, errors can sometimes occur, resulting in unintended modifications in the genome, known as "off-target effects". This lack of precision can be concerning, especially when editing human cells.

- Efficiency: The efficiency of genetic editing can vary depending on the type of cells and organism targeted. Some cells may be more difficult to modify than others, and efficiency may depend on factors such as the delivery capability of genetic editing tools into cells.

- Size of insertions: Inserting larger genetic sequences may be technically challenging, limiting the ability to insert large portions of DNA into the genome.

- Challenges in the context of gene therapies: In the context of gene therapies, delivering the genetic editing tool to all necessary cells can be challenging, especially for deep tissues or inaccessible organs.

- Genomic stability: Modifications to the genome may not be stable in the long term, and edited cells may undergo undesirable mutations over time.

❖ Possibilities of Cognitive Modification

Different Types of Envisaged Genetic Modifications

Prevention of Neurological Diseases

One of the most promising applications of genetic editing in the brain is the prevention of neurological diseases. By modifying genes associated with conditions such as Alzheimer's disease or Parkinson's disease, it could be possible to mitigate genetic risks.

Enhancement of Cognitive Abilities

Genetic editing could also be used to specifically enhance cognitive abilities. This could include boosting memory, concentration, and other key mental functions by targeting genes involved in these processes.

Memory Enhancement

Memory, a fundamental aspect of cognitive functions, could be a key target for cognitive enhancement. Genetic interventions to strengthen memory formation and retention could have significant implications in areas such as learning and productivity.

Development of Specific Cognitive Skills

Cognitive modification could also be aimed at developing specific cognitive skills. For example, by promoting neural plasticity in certain brain regions, one could increase the ability to quickly learn new skills.

Modulation of Emotions and Behavior

A more controversial approach would be the modulation of emotions and behavior through genetic editing. This raises complex ethical questions related to the manipulation of behavioral traits and the definition of what is considered "normal" or "enhanced".

Genes Related to Intelligence and Cognitive Abilities

Recent research has identified specific genes associated with intelligence and cognitive abilities. Genetic editing could allow for modulation of the expression of these genes to enhance traits such as memory, learning, and problem-solving.

Studying genes related to intelligence and cognitive abilities is a complex endeavor aimed at deciphering the genetic foundations of human intelligence. Research has identified several genes that appear to play a role in cognitive development, although the genetic landscape of intelligence is infinitely complex and far from fully understood.

Recognized as an evolving subject, research on specific genes related to intelligence and cognitive abilities has yielded fascinating yet intricate discoveries. Studies have revealed several genes that seem to influence higher brain functions. For example, the CHRM2 gene has been associated with memory and learning, while the COMT gene has been linked to thinking processes and problem-solving. Additionally, the FADS2 gene has been identified as impacting working memory.

The protein associated with the NRXN1 gene, which plays a role in synaptic formation, has also been linked to superior cognitive performance. Variations in the KIBRA gene have been associated with episodic memory, highlighting the importance of this gene in the ability to recall specific events. Furthermore, studies have shown that the BDNF gene, involved in neuronal growth and survival, is associated with enhanced cognitive functions.

However, it is crucial to note that these discoveries represent only a fraction of the complex genetic panorama of intelligence. Many other genes and their complex interactions come into play, and the relative influence of these genes may vary from person to person. Additionally, the environment, including factors such as education and exposure to intellectually stimulating stimuli, plays a crucial role in the manifestation of cognitive abilities. Research on specific genes related to intelligence continues to evolve, offering increasingly precise, albeit nuanced, insights into the genetic bases of our mental faculties.

Genes Related to Neurotransmission

Genetic editing can also influence neurotransmission, the process by which neurons communicate with each other. By modifying genes involved in the production, release, or reception of neurotransmitters, it could be possible to optimize neuronal circuits to improve cognitive function.

Specific genes that regulate the production, release, or reception of neurotransmitters play a fundamental role in neuronal communication and nervous system functions. Some of these genes are crucial for the proper functioning of the brain and have been associated with neurological and psychiatric conditions.

- Neurotransmitter synthesis genes: Enzymes responsible for neurotransmitter synthesis are governed by specific genes. For example, the TH gene codes for tyrosine hydroxylase, a key enzyme in dopamine synthesis, while the TPH gene regulates serotonin synthesis.
- Neurotransmitter transporter genes: Neurotransmitter transporters, such as the SLC6A4 gene involved in serotonin transport, are essential for recycling neurotransmitters after their release into the synapse.
- Neurotransmitter receptor genes: Neurotransmitter receptors, such as dopamine receptors coded by genes like DRD1 and DRD2, are crucial for the transmission of chemical signals between nerve cells.
- Genes related to neurotransmitter degradation: Enzymes such as monoamine oxidase (MAO), regulated by the MAOA and MAOB genes, participate in neurotransmitter degradation, thus influencing their availability in the synapse.

- Genes involved in neurotransmitter release: Proteins like synapsin, coded by the SYN1 gene, play a role in regulating neurotransmitter release at synapses.

Genetic variations in these genes can contribute to individual differences in the functioning of the nervous system, influencing traits such as behavior, cognition, and susceptibility to neurological or psychiatric disorders. Research on these specific genes provides valuable insights into the molecular mechanisms underlying neurotransmitter regulation and opens perspectives for understanding neurological and psychiatric disorders.

Scientific Limitations of Cognitive Modification

Genetic cognitive modification, while fascinating in its potential, raises a number of scientific limitations and realities that require careful consideration. Here are some of the main concerns:

- Cognitive complexity: Human cognition is a complex trait resulting from the interaction of many genes, environmental factors, and life experiences. Genetically modifying a single gene may not suffice to significantly influence complex cognitive abilities such as intelligence.

- Limited scientific knowledge: Current understanding of the genetic mechanisms underlying cognition is still incomplete. Modifying these mechanisms without a thorough understanding of the potential consequences can lead to unpredictable outcomes.

- Genetic heterogeneity: Genetic diversity among individuals is vast. A one-size-fits-all approach to cognitive genetic modification cannot account for this variability, and what works for one person may not work the same way for another.

- Side effects and unforeseen consequences: Genetic modification can lead to undesirable side effects, including unintended changes in the genome (off-target effects). These effects can have serious health consequences.

- Brain development: Cognition is closely linked to brain development, an extremely complex and delicate process. Any genetic intervention in this area must take into account critical stages of development.

- Gene-environment interaction: Genetic effects on cognition can interact in complex ways with environmental factors. Ignoring these interactions may underestimate the influence of the environment on cognitive outcomes.

In summary, while genetic cognitive modification is of considerable interest, it faces substantial challenges, both scientific and ethical. In-depth research, increased understanding of genetic mechanisms, and rigorous ethical consideration are essential before practical applications in this field can be envisaged.

❖ Medical and Therapeutic Applications

Neurological Disorders

Neurodegenerative Diseases

Brain genetic editing holds promise in the treatment of neurodegenerative diseases such as Parkinson's and Alzheimer's. Researchers are exploring the possibility of correcting genetic abnormalities responsible for these conditions, or even replacing defective cells with genetically modified cells.

Genetic Brain Development Disorders

Genetic brain development disorders, such as autism, could also benefit from genetic editing. By modifying the genes involved in these disorders, it may be possible to potentially alleviate symptoms and improve the quality of life for affected individuals.

Personalized and Precise Treatments

Genetic editing would enable more personalized and precise treatments. By specifically targeting genes associated with each patient, interventions could be tailored based on individual genetic characteristics, thereby increasing treatment effectiveness.

Possibilities for Mental Illness Prevention

Preventive Approaches for Psychiatric Disorders

Genetic editing offers the possibility to develop preventive approaches for psychiatric disorders such as depression, anxiety, and schizophrenia. By identifying genetic factors predisposing to these conditions, it might be conceivable to correct them before symptoms even arise.

Early Intervention and Genetic Risk Modification

Early identification of genetic risk factors would allow for early intervention, altering the potential course of mental illnesses. However, this raises ethical questions regarding intervening in a person's life before they even develop a mental condition.

Balance Between Prevention and Autonomy

Preventing mental illnesses through genetic editing raises delicate questions about the balance between preventing mental suffering and respecting individual autonomy. To what extent is it ethical to genetically modify a person to prevent conditions that might arise in the future?

❖ Ethical Dilemmas

Ethical Questions

Self-Determination and Consent

One of the main ethical dilemmas of brain genetic editing lies in respecting individual self-determination. The issue of informed consent becomes crucial when considering genetic modification that can influence aspects as intimate as cognitive and emotional abilities.

Risks of Social Inequalities

The possibility of uneven use of genetic technology raises major concerns related to social justice. If access to these interventions is limited based on factors such as social class, it could intensify existing disparities.

Definition of "Normality"

The redefinition of cognitive and behavioral norms also poses ethical challenges. Who determines what is considered "normal" and "enhanced"? This question raises concerns about the stigmatization of natural variations in cognition.

Confidentiality and Self-Determination

Privacy Risks

Brain genetic editing raises major concerns regarding genetic privacy. Information about brain modification could be sensitive and potentially misused, jeopardizing individuals' privacy.

Self-Determination and Personal Identity

Genetic brain modification raises profound questions about self-determination and personal identity. To what extent can a person direct their own cognitive development without compromising their authenticity and integrity as a unique individual?

Informed Consent and Education

Ensuring informed consent requires thorough education on the risks, benefits, and long-term implications of brain genetic editing. This raises the question of whether individuals are currently sufficiently informed to make informed decisions about such complex modifications.

Potential Risks for Genetic Diversity and Society

Loss of Genetic Diversity

Brain genetic editing could potentially lead to a loss of genetic diversity. If certain cognitive or behavioral traits become predominant due to widespread genetic modifications, it could decrease natural diversity within the population.

Effects on Social Dynamics

Changes resulting from brain genetic editing could have profound impacts on social dynamics. Cognitive inequalities could create tensions between different segments of society, leading to social and economic challenges.

Impacts on Distributive Justice

Access to brain genetic editing raises crucial questions of distributive justice. Who has the right to access these technologies and how are resources distributed fairly? These questions are essential to avoid a concentration of power in the hands of a few.

❖ Regulatory and Normative Framework

Need for Regulation

Avoiding Misuse and Abuse

The power of brain genetic editing raises concerns about its responsible use. Strict regulation is necessary to prevent ethical lapses and potential abuses, ensuring that this technology is used for the common good.

Intervention Safety and Individual Protection

Regulation is essential to ensure the safety of brain genetic editing interventions. Potential risks to health, both individual and collective, must be thoroughly assessed. Protecting the rights and dignity of individuals is a priority.

Research and Clinical Trial Oversight

Research in the field of brain genetic editing requires close monitoring to ensure ethical protocols and the reliability of results. Clinical trials must adhere to strict standards to evaluate the effectiveness and safety of interventions.

Role of Governments and International Organizations

National Legislation and International Coordination

Governments play a crucial role in establishing strong national legislation regarding brain genetic editing. However, international coordination is necessary to avoid discrepancies and ensure consistent application of standards globally.

Ongoing Monitoring and Adaptability

International organizations such as the WHO and UNESCO must play a central role in ongoing monitoring and adaptability of regulatory standards. Rapid advancements in this field demand constant monitoring and the ability to adjust regulations quickly.

Collaboration with Society and Experts

Regulation of brain genetic editing must also involve civil society and multidisciplinary experts. Transparency, public participation, and open dialogue are essential to ensure that regulations meet the needs and values of society.

❖ Future Perspectives

Next Steps in Brain Genetic Editing

Refinement of Genetic Editing Techniques

Next steps will likely include refining brain genetic editing techniques. More precise, less invasive, and more effective methods will be developed, allowing for more specific and sophisticated genetic manipulation.

Exploration of New Genes and Targets

Research will focus on exploring new genes related to the brain and identifying new targets for genetic editing. Further understanding the complex genetics of the brain will open up new intervention possibilities.

Expanded Applications to Other Neurological Conditions

As knowledge advances, brain genetic editing could be extended to other neurological conditions, offering potential solutions for a wide range of disorders.

Long-Term Implications for Humanity

Human Evolution and Genetic Diversity

Brain genetic editing could have long-term implications for human evolution and genetic diversity. Genetic manipulation could influence the frequency of certain genetic traits in the population, raising questions about diversity and adaptability of the human species.

Transformation of Cognitive Abilities and Performance

Enhancing cognitive abilities through genetic editing could transform how individuals think, learn, and problem-solve. This will raise questions about equity in access to these cognitive enhancements and how it might affect society.

New Ethical and Philosophical Frontiers

The long-term implications of brain genetic editing will open new ethical and philosophical frontiers. Questions about the definition of human identity, normality, and the intrinsic value of the individual will emerge, requiring deep reflection and ongoing ethical dialogue.

Impact on Humanity

Mastering Evolutionary Forces

Brain genetic editing opens the door to the possibility of mastering the evolutionary forces that have shaped humanity over millennia. Rather than relying solely on natural mechanisms of natural selection, genetic editing offers the ability to deliberately guide human evolution.

Targeted Interventions on Cognitive Traits

Interventions could be targeted at specific traits such as intelligence, memory, or emotional resilience. The possibility of deliberately shaping these characteristics raises fundamental questions about the nature and direction of human evolution.

Accelerated Evolution

Genetic editing could accelerate the evolutionary process, allowing for significant changes in a much shorter period than natural selection would. This raises questions about the stability and adaptability of newly genetically modified traits.

Genetic Convergence

Genetic editing could potentially lead to genetic convergence, where specific traits become predominant at the expense of genetic diversity. This could have profound implications for humanity's ability to adapt to changing environments.

Risks of Cognitive-Genetic Uniformity

The pursuit of cognitive enhancement via genetic editing poses the risk of creating cognitive-genetic uniformity. If specific traits become universally valued, it could lead to a loss of cognitive and behavioral diversity within the population.

Balance Between Homogeneity and Adaptability

Managing human evolution through genetic editing requires a delicate balance between preserving genetic homogeneity for specific benefits and maintaining genetic diversity that provides adaptability to unforeseen changes.

4

Into the Inner Infinite: The Mysteries of Consciousness

Exploring the Human Consciousness

This chapter delves into the depths of human consciousness, exploring its neurobiological foundations, altered states, and its complex interaction with perception and self-awareness.

❖ Nature of Consciousness

Consciousness, this complex and mysterious phenomenon, remains at the very heart of human experience. Through defining and exploring its characteristics, it becomes possible to grasp the fundamental importance of consciousness in our existence.

Definition of Consciousness

Consciousness, often described as an elusive phenomenon, is intrinsically linked to perception and subjective experience. It refers to the ability to perceive and understand one's environment, thoughts, emotions, and sensations. Consciousness encompasses a variety of dimensions, from simple sensory awareness to reflexive and metacognitive consciousness.

At a fundamental level, consciousness allows for a unified experience of the world, integrating sensory information into a coherent perception. It is the thread that weaves together the different aspects of our inner and outer reality.

Consciousness vs Cognition

Consciousness and cognition are two interconnected but distinct concepts that play a crucial role in understanding the human mind. Here's how they differ:

Consciousness

Consciousness refers to the state of being awake and aware of oneself and the environment. It is the subjective experience of being conscious and alert. It allows us to have a subjective experience of the world and ourselves.

Consciousness can be divided into several aspects, including:

1. Perceptual Consciousness: The ability to perceive and interpret sensory stimuli from the environment.

2. Reflexive Consciousness: The ability to reflect on oneself, one's thoughts, and experiences.

3. Self-Consciousness: Awareness of one's own mental state, emotions, and existence as a distinct individual.

Cognition

Cognition encompasses a set of mental processes related to the acquisition, storage, manipulation, and use of information. It is often considered the underlying process by which we analyze data.

The main domains of cognition include:

1. Perception: How we interpret sensory information to perceive the world around us.

2. Memory: The process of storing and retrieving information.

3. Problem-solving: Mental processes related to problem-solving and decision-making.

4. Language: The ability to use and understand language for communication.

5. Attention: The ability to focus on specific information while ignoring other stimuli.

Characteristics of Consciousness

Consciousness is characterized by several distinctive elements that define its profound nature. Firstly, it is intentional, meaning it is always directed towards something. Whether it's a thought, a sensation, or an external object, consciousness always has an object.

Additionally, consciousness is selective. Although we are constantly exposed to a multitude of stimuli, consciousness filters and selects certain information for in-depth processing. This selectivity contributes to our ability to focus our attention on specific aspects of our experience.

Consciousness is also dynamic and ever-changing. It continuously evolves in response to fluctuations in our environment and mental state. Sleep cycles, dreams, meditative states, and emotional fluctuations are all manifestations of this inherent dynamism of consciousness.

Fundamental Importance of Consciousness

Consciousness is crucial in the human experience. It is the foundation upon which our ability to make sense of our reality, interact with the surrounding world, and understand our own existence rests.

• Reality Construction: Consciousness plays a central role in constructing our subjective reality. It filters, interprets, and gives meaning to sensory information, thus creating our individual experience of the world.

• Social Interaction: Consciousness is crucial for our social interaction. It allows us to understand the intentions, emotions, and thoughts of others, thereby fostering communication and cooperation.

• Inner Reflection: Consciousness also offers the opportunity for inner reflection. It allows us to explore our own thoughts, emotions, and motivations, facilitating the process of self-reflection and personal development.

• Adaptation to Change: Consciousness plays an essential role in our ability to adapt to a constantly changing environment. It enables us to learn, remember, and adjust based on our past experiences.

❖ Neurobiological Foundations

Consciousness finds its foundations in sophisticated neurobiological mechanisms that orchestrate brain activity.

Brain Mechanisms of Consciousness

Attention Regulation Systems

The neurobiological mechanisms that regulate attention are crucial for consciousness. Thalamic nuclei, often referred to as the "gateway to consciousness," filter and relay sensory information to the cerebral cortex. The prefrontal cortex, located at the front of the brain, plays a crucial role in selecting relevant stimuli and regulating attention. It is within these complex interactions between the thalamus and the prefrontal cortex that the basis of consciousness is formed.

Information Integration

The integration of information from different parts of the brain is a key element of consciousness. The parietal cortex, which integrates sensory and spatial information, as well as the temporal cortex, involved in processing auditory and visual stimuli, are major players in this integration. Close connections between these regions allow for the creation of a unified and coherent representation of the environment.

Brain Networks

Neuroscientists have identified specific brain networks associated with particular aspects of consciousness. The default mode network, including the posterior cingulate cortex and the medial prefrontal cortex, is active when the mind is in a state of rest or self-reflection. In contrast, the attention network, involving the parietal cortex and the frontal cortex, is engaged during tasks that require concentration and focus of attention.

These networks interact dynamically to support the various facets of consciousness.

Neuronal Activation

Neuronal activation, particularly in the cortex, is a fundamental component of consciousness. Cortical neurons communicate through electrochemical impulses, creating complex patterns of activity. Neuronal oscillations, characterized by changes in the frequency and amplitude of electrical signals, are also associated with different states of consciousness, from alert wakefulness to deep sleep.

Brain Plasticity

Brain plasticity also significantly contributes to consciousness. Experiences, learning, and even meditation can alter the structure and function of the brain, thus influencing consciousness.

Recent Advances in Consciousness Research

Neuroscience and Brain Imaging

Advancements in brain imaging technologies, such as fMRI and EEG, allow researchers to observe brain activity with increasing resolution. This leads to a better understanding of the brain regions involved in consciousness and the patterns of interaction between different brain regions.

Artificial Intelligence and Computational Models

Artificial intelligence has become a powerful tool in the study of consciousness. Researchers use computational models to simulate and test different theories of consciousness. These models help to better understand how brain processes could lead to conscious experience.

Studies on Altered States of Consciousness

Research on altered states of consciousness, such as those induced by meditation, psychoactive substances, or pathological states, has gained importance. They offer unique insights into the nature of consciousness and how it can be altered.

Integration of Cognitive Sciences and Phenomenology

Researchers are increasingly seeking to combine scientific approaches with phenomenological methods to obtain a more holistic understanding of consciousness. This involves considering both the objective and subjective aspects of conscious experience.

Key Brain Regions

Prefrontal Cortex

The prefrontal cortex, often called the seat of intelligence, plays a central role in consciousness. This region is involved in decision-making, planning, working memory, and emotional regulation. Lesions in the prefrontal cortex can lead to significant alterations in consciousness, affecting the ability to devise strategies, maintain attention, and interact socially.

Parietal Cortex

The parietal cortex contributes to the spatial representation of the body and sensorimotor perception. It is crucial for the construction of bodily awareness and understanding of the body's position in space. Studies show that alterations in the parietal cortex can lead to altered experiences of body consciousness, such as disembodiment, where the individual feels detached from their own body.

Thalamus

The thalamus is a subcortical structure that plays a crucial relay role in transmitting sensory signals to the cortex. Thalamic lesions can lead to alterations in vigilance and perception, highlighting its central role in consciousness construction.

Anterior Cingulate Cortex

The anterior cingulate cortex is involved in emotional regulation and self-awareness. It plays an essential role in managing cognitive and emotional conflicts, thus contributing to the stability of consciousness. Alterations in this region may be associated with mood disorders and self-perception issues.

❖ Theories of Consciousness

Several theories have been proposed to try to explain consciousness, each offering a different perspective on this fascinating phenomenon. Here are some of the most influential theories.

Integrated Information Theory

Integrated Information Theory (IIT) was developed by neuroscientist Giulio Tononi to attempt to answer the complex question of what distinguishes conscious systems from non-conscious ones.

Here are some key principles of IIT:

1. Axiom of integrated information: According to IIT, a quantifiable measure called Φ (phi) characterizes the level of integration of a system. The higher Φ is, the more conscious the system is.

2. Causal exclusion: A conscious system must have components that are strongly interconnected so that information cannot be divided into independent parts. If part of the system is isolated, it decreases integrated information and, consequently, consciousness.

3. Specific informative character: IIT emphasizes the specific nature of information, meaning that the integrated information in a conscious system is specific to that system and cannot be reduced to the information available in its individual parts.

4. Expansion of the repertoire of possible states: A conscious system is capable of existing in many different states, which Tononi calls the "repertoire of possible states." The broader the repertoire, the more conscious the system is.

IIT has garnered considerable interest in the scientific community due to its attempt to provide a mathematical theory of consciousness.

Neuronal Correlates Theory

Francis Crick, best known for his discovery of the structure of DNA with James Watson, also became interested in the question of consciousness in the later years of his life. Along with neuroscientist Christof Koch, Crick developed the theory of Neuronal Correlates of Consciousness (NCC).

This theory posits that consciousness emerges from specific neuronal activities in the brain. Thus, the central idea is that certain configurations and neuronal processes are directly related to conscious experience. These neuronal correlates may include patterns of synchronization or binding between neurons in different brain regions.

The NCC theory suggests that for a state to be conscious, there must be synchronized neuronal activity in associative cortical regions, where the integration of information from different parts of the brain occurs. Thus, the neuronal correlates of consciousness are these specific patterns of neuronal activity that coincide with conscious experience.

Complex Systems Theory

Complex systems theory is an approach that can be applied to various scientific domains. The fundamental idea behind this theory is that complex phenomena emerge from the interaction and organization of many interconnected elements.

Regarding consciousness, some researchers apply complex systems theory to understand how the many elements of the brain interact to produce conscious experience. Instead of focusing solely on specific brain regions, this approach examines the overall dynamics of the brain network.

Here are some key points associated with applying complex systems theory to consciousness:

1. Emergence: Complex systems theory is particularly interested in emergent phenomena, i.e., how new and often unpredictable properties can arise from the collective functioning of the constituent elements of a system.

2. Nonlinear Dynamics: Complex systems often involve nonlinear dynamics, meaning that small changes in one part of the system can lead to significant and sometimes unpredictable changes in the entire system.

3. Adaptation: Complex systems have the ability to adapt to changes in their environment. In the context of consciousness, this could mean that brain dynamics adapt to produce conscious experiences in response to external and internal stimuli.

4. Connectivity: Connectivity between different parts of the brain is a key aspect of complex systems theory. The emergence of consciousness is often associated with specific patterns of neuronal connectivity and how information flows through the brain network.

5. Self-organization: Complex systems have the ability to self-organize, meaning they can evolve into ordered states without direct external control.

Theory of Prediction

The Theory of Prediction is a perspective in cognitive neuroscience that proposes the brain functions as a prediction system. This theory falls within the broader framework of the Bayesian view of the brain and seeks to explain how the brain generates perceptions by actively anticipating sensory stimuli.

Here are some key principles of this theory:

1. Prediction Principle: According to this theory, the brain constantly generates predictions about the external world. These predictions are based on internal models constructed from past experiences.

2. Prediction Errors: When the brain's predictions do not match actual sensory stimuli, a "prediction error" occurs. The brain then adjusts its internal models to minimize these errors. This contributes to optimizing perception and understanding of the world.

3. Ascending and Descending Propagation: Information flows both bottom-up (from sensory perception to higher levels of processing) and top-down (from higher brain regions to more sensory regions). Predictions descending from internal models influence how sensory stimuli are perceived.

4. Bayesian Inference: The prediction theory fits within a framework of Bayesian inference, which uses probabilities to update the brain's beliefs based on new information. The brain continually adjusts its hypotheses about the world based on the gap between predictions and real experiences.

The Theory of Prediction suggests that consciousness itself may emerge from this process of prediction and error correction.

Some researchers suggest that consciousness is linked to how the brain actively processes information, by predicting and continually adjusting its internal representations based on sensory inputs.

This theory has implications in various domains of cognitive neuroscience, psychology, and artificial intelligence, and it continues to garner significant interest in understanding the underlying mechanisms of perception and consciousness.

Enactive Theory

The Enactive Theory is a perspective in cognitive science that proposes a radically different approach to understanding cognition and consciousness. This theory was primarily developed by philosopher Francisco Varela, biologist Humberto Maturana, and psychologist Eleanor Rosch.

The enactive theory focuses on the idea that sensorimotor activity and the organism's interaction with its environment are essential for understanding cognition and consciousness. Unlike some traditional cognitive theories that emphasize mental representation, enaction suggests that the mind inherently emerges from organism-environment interaction.

Here are some key principles of this theory:

1. Incarnation: Enaction emphasizes the concept of incarnation, highlighting that the mind is closely connected to the body and its interaction with the environment. Cognitive and conscious processes are rooted in bodily and sensory experience.

2. Action and Perception: According to this theory, action and perception are intrinsically linked. Organism

actions are not simply responses to sensory signals but rather integral parts of the cognitive process.

3. Organism-Environment Circularities: Enaction emphasizes the organism-environment circularity, where the organism and its environment are considered co-defining. Organism actions help define its environment, and vice versa.

4. Autonomy: Autonomy is a central concept in enaction. Autonomous cognitive systems are capable of self-organizing and maintaining their internal coherence during interaction with their environment.

The theory of enaction has profound implications for understanding consciousness by emphasizing sensorimotor activity and dynamic interaction with the environment as the foundation of cognition. It has influenced various fields, including psychology, neuroscience, and robotics, by changing how we think about the nature of the mind and consciousness.

Theory of Global Information Attention

This theory suggests that consciousness emerges when attention is globally directed towards certain information in the brain. Focusing attention would be crucial for conscious experience.

Thus, focusing attention on certain aspects of the environment or experience is often seen as a key mechanism for certain information to access consciousness.

A common approach to addressing these issues is to consider attention as a gateway between sensory stimuli and consciousness. When attention is directed towards a specific part of the environment or information, that information is more likely to become conscious.

❖ Self-Awareness and Perception

Self-awareness, the internal mirror allowing us to perceive ourselves as distinct individuals, is closely linked with our perceptual processes.

Interactions Between Consciousness and Perceptual Processes

Perceptual Selectivity

Perceptual processes, encompassing the reception and interpretation of sensory information, play a crucial role in constructing self-awareness. Perceptual selectivity, the ability to attend to certain stimuli over others, directly influences our self-awareness. Sensory experiences shape how we perceive ourselves and how we are perceived by others.

Sensorial Integration

Sensorial integration, the process by which the brain combines information from different senses, is essential for self-awareness. The coherent image we have of our body and identity results from the harmonious integration of visual, auditory, tactile, and proprioceptive stimuli. Disruptions in this process can lead to alterations in self-awareness, as observed in conditions such as schizophrenia.

Self-Recognition

Self-awareness also includes the ability to recognize oneself, whether in a mirror or through other sensory modalities. This ability involves brain regions such as the prefrontal cortex and the anterior cingulate gyrus.

Role of Senses in Self-Consciousness Construction

Vision

Vision is often considered the dominant sense in constructing self-awareness. Visual perception of our body, facial expressions, and immediate environment profoundly influences our identity. Studies show that visual alterations, such as those induced by distortion glasses, can temporarily modify self-perception.

Hearing

Although less directly related to self-awareness, hearing plays an essential role in how we perceive ourselves socially. Voice, tones, and modulations influence our understanding of our own identity and how we are perceived by others.

Touch and Proprioception

Touch, proprioception (perception of body position and movement), and tactile sensations contribute to our bodily awareness. How we feel our body in space and interact with our environment plays a fundamental role in constructing our physical identity.

Olfaction and Taste

Although often underestimated, olfaction and taste also influence our self-awareness. Familiar smells can trigger memories related to our identity, while taste can be associated with cultural and personal experiences that shape our self-perception.

Exploration of Self-Recognition Ability

Researchers use various methods, including mirror and photography tests, to probe how individuals perceive and recognize their own image.

The mirror test, classically used in humans and some animals, evaluates whether a subject interacts with their reflection in a way indicating self-recognition, often observed through specific behaviors such as touching parts of the body normally invisible.

Photography experiments, on the other hand, may include digital manipulations of the self-image to examine how individuals react to visual alterations.

These explorations offer insights into identity construction and self-awareness, aiding in better understanding how these processes develop, particularly through cognitive developmental stages and considering cultural factors.

Psychological and Social Implications

Impact on Self-Esteem

How we perceive our own body and identity directly influences our self-esteem. Social norms, aesthetic ideals, and social comparisons play a crucial role in constructing this self-perception. Distortions in self-perception can contribute to issues such as eating disorders and body dysmorphia.

Interpersonal Relationships

Self-awareness also shapes our interpersonal relationships. The perception we have of our own identity influences how we present ourselves to others, how we form social bonds, and how we interact in the world.

Cultural Identity

Self-awareness is deeply linked to our cultural identity. Cultural norms, expectations, and values shape how we perceive ourselves as individuals within the context of our specific culture. Cultural variations in self-awareness construction highlight the diversity of this experience.

Psychopathological Consequences

Alterations in self-awareness can be associated with various psychopathological disorders. For example, in depression, self-perception can be tinged with pessimism and excessive self-criticism. Conversely, in some personality disorders, there may be a marked distortion of self-image.

❖ Alterations of Consciousness

Ordinary consciousness, often referred to as the state of daily consciousness, characterizes most of our waking life. It is the state in which we are aware of our environment, thoughts, emotions, and actions coherently. It is the substrate of our daily reality, where the perception of time, space, and self is relatively stable.

On the other hand, altered consciousness refers to states in which perception and experience differ from the normal. This can include states of trance, mystical ecstasy, hypnosis, lucid dreaming, or the influence of psychoactive substances. These altered states can modify the perception of time, self, and reality, often offering unusual and sometimes profound perspectives.

Alterations of Consciousness: Between Shadow and Light of the Mind

The study of altered states of consciousness offers a fascinating window into the depths of the human mind. These alterations reveal dimensions of consciousness that often escape our everyday understanding.

Alterations of Consciousness Related to Mental Illness

- Depression: Depression can profoundly alter self-awareness and the world. Pessimistic thoughts, loss of interest in once-enjoyed activities, and a negative view of oneself can cloud the perception of the world. Consciousness becomes tinged with persistent sadness and despair, affecting how a person interacts with their environment.

- Schizophrenia: In schizophrenia, alterations in consciousness can manifest as hallucinations, disorganized thoughts, and a disintegration of reality. Self-perception and perception of the world can be profoundly disturbed, often marked by confusion between reality and fantasy.

- Anxiety Disorders: Anxiety disorders can alter consciousness by generating obsessive thoughts, negative anticipations, and mental hyperactivity. The perception of the world often becomes filtered through anxiety, creating a subjective reality marked by fear and apprehension.

Alterations Related to Psychoactive Substances

- Drug Effects: Psychoactive substances, whether legal or illicit, can induce spectacular alterations in consciousness. Hallucinogens, such as LSD, can lead to sensory distortions and dissociation from reality. Opioids, on the other hand, can numb consciousness, creating a state of lethargy and euphoria.

- Alcohol and Cognitive Impairment: Alcohol, though socially accepted, is a substance that significantly alters consciousness. It can numb cognitive functions, affect motor coordination, and alter judgment. These alterations often contribute to changes in self-perception and impulsive behaviors.

Alterations by Specific Phenomena

- Meditation: Meditative states, such as those attained through deep meditation, can alter consciousness in unique ways. Meditators sometimes experience a dissolution of boundaries between self and the world, increased mental clarity, and a reduction in compulsive mental activity. These alterations are often associated with a perception of the world imbued with calmness and connectivity.

- Lucid Dreams: Lucid dreams, where the individual becomes aware they are dreaming while still dreaming, represent an intriguing form of altered consciousness during sleep. In these moments, self-perception and perception of the world can be shaped by unbridled creativity and the infinite possibilities of lucid dreaming.

- Near-Death Experiences (NDEs): NDEs offer unique insights into alterations of consciousness. These experiences may involve sensations of out-of-body experiences, passing through a tunnel, and encounters with spiritual entities. Changes in self-perception and perception of the world during these experiences raise profound questions about the nature of consciousness and reality.

Implications for Understanding Consciousness

Plastic Nature of Consciousness

The study of alterations in consciousness reveals the plastic nature of the human mind. Consciousness is not a static entity but rather a constantly evolving reality, capable of adapting to extraordinary conditions or distorting under the influence of mental disorders.

Understanding the Boundaries of Reality

Alterations in consciousness push the boundaries of subjective reality. They question our usual perceptions of self and the world, showing that consciousness can be fluid and malleable, subject to extraordinary variations.

Reflection on the Nature of Reality

Meditative experiences, lucid dreams, and NDEs invite deep reflection on the nature of reality. They raise questions about the relationship between consciousness, the brain, and objective reality, paving the way for philosophical and scientific explorations.

❖ Artificial Consciousness

The creation of artificial consciousness represents the pinnacle of humanity's technological aspirations, opening the door to possibilities that defy imagination.

Current State of Artificial Consciousness Creation

Artificial Intelligence

Current research in the field of artificial consciousness creation is primarily focused on the development of artificial intelligence (AI). Current AI systems can learn from data, solve complex problems, and perform tasks that previously required human intervention. However, current AI is not conscious in the true sense of the term.

Simulation of Consciousness

Some researchers explore approaches to simulate aspects of consciousness through artificial neural networks that attempt to reproduce certain characteristics of human cognitive processes, but there remains a fundamental difference between simulation and true consciousness.

Brain-Machine Interfaces

Brain-machine interfaces are another frontier of research. These interfaces aim to establish a direct connection between the human brain and machines, enabling bidirectional information exchange. While this may enhance communication between humans and machines, the creation of complete artificial consciousness remains a distant prospect.

Future Horizons of Artificial Consciousness

The future horizons of artificial consciousness involve a deeper understanding of the mechanisms of human consciousness. Advances in neurobiology and understanding of human cognitive processes could illuminate the path to creating artificial consciousness by replicating these mechanisms.

Some experts also envision the gradual development of synthetic consciousness, where machines could acquire increasing levels of autonomy and consciousness, although these levels remain far from equaling the complexity of human consciousness.

Furthermore, the idea of closer human-machine fusion is also explored. Progress in this area could allow for smoother interaction between the human mind and computer systems, creating a symbiosis that might resemble a form of shared consciousness.

Debates Surrounding Artificial Consciousness

Rights of Artificial Conscious Entities

One of the central ethical debates concerns the rights of artificial conscious entities. If a machine were endowed with a form of consciousness, should it be entitled to rights equivalent to those of human beings? This question raises complex philosophical and legal challenges.

Responsibility and Decision-Making

The issue of responsibility and decision-making in the context of artificial consciousness is crucial. If a machine were to make autonomous decisions, how would responsibility be attributed in case of problematic or harmful actions? This debate is particularly critical with the increasing deployment of AI in fields such as healthcare, finance, and security.

Impact on Employment

Increased automation through AI raises concerns about its impact on employment. If machines were to acquire consciousness and autonomy, it could question the place of humans in the workforce, with significant economic and social consequences.

❖ Collective Consciousness

Collective consciousness is a fascinating concept referring to shared understanding, beliefs, and values that emerge within a group, transcending individuals to create a collective cognitive entity. It plays a central role in sociology and social psychology.

Definition of Collective Consciousness

Collective consciousness is a concept developed by the French sociologist Émile Durkheim (1858-1917). Durkheim played a major role in establishing sociology as a distinct academic discipline and is also known for his pioneering studies on suicide, where he demonstrated how social factors could influence individual behaviors.

According to Durkheim, collective consciousness refers to the set of beliefs, values, norms, and attitudes shared by members of a society. It is a social force that unites individuals by providing them with a set of common references and creating a shared social identity. Collective consciousness is distinct from individual consciousnesses and influences the behavior of members of society. It manifests through institutions, rituals, traditions, and other forms of social life.

Emergence of Collective Consciousness

Interconnection of Minds

Collective consciousness arises from the interconnection of individual minds. Each individual, carrying their unique experiences, beliefs, and perceptions, contributes to the fabric of this collective consciousness. It is within this diversity of individuals that the richness of collective consciousness lies, a complex symphony of ideas, values, and emotions.

Reciprocal Influence

Individual consciousness and collective consciousness continuously influence each other. Ideas shared within a group can shape individual perception, just as individual convictions contribute to the construction of collective reality. This dynamic interaction creates a delicate balance between individual autonomy and social cohesion.

Social Construction of Reality

The social construction of reality is a fundamental concept in sociology, elaborated notably by sociologists Peter L. Berger and Thomas Luckmann in their eponymous work published in 1966. This perspective argues that reality is not simply an objective entity existing independently of individuals but rather that it is socially constructed through interactions and social processes. Thus, reality is shaped by shared beliefs, norms, values, and institutions of a society. Individuals learn to perceive and interpret the world around them through the prism of these social constructions. Language, symbols, rituals, and institutions are mechanisms through which reality is defined and maintained.

Sociological Implications

By forging a shared identity through beliefs, values, and norms, collective consciousness creates strong social cohesion. It acts as a social cement that promotes solidarity and cooperation, contributing to the stability of society. It also plays a crucial role in regulating individual behaviors, establishing norms that define acceptable boundaries in a given society. Additionally, collective consciousness influences how individuals internalize cultural principles.

However, tensions can emerge between individual and collective consciousness, raising questions about individual freedom and conflicts of values within a community. Moreover, collective consciousness is not static. It is subject to changes and conflicts that arise when groups with divergent consciousnesses come into contact. Social movements, revolutions, and political changes often result from tension between different expressions of collective consciousness.

Cultural Implications

Cultural Transmission

Collective consciousness acts as an essential vector for cultural transmission. Myths, traditions, and shared values within a society are perpetuated by collective consciousness. It becomes the receptacle where knowledge and ideals are preserved and passed down from generation to generation.

Cultural Evolution

The evolution of collective consciousness reflects cultural evolution. Changes in social attitudes, moral norms, and aesthetic perceptions are indicators of the dynamics of collective consciousness over time. Artistic movements, technological advances, and political changes are reflections of this evolution.

Cultural Diversity

Collective consciousness is not universal but rather specific to each culture. Cultural diversity is highlighted by the different ways in which social groups apprehend the world, attribute meaning to existence, and interact with one another. This diversity enriches the overall mosaic of global collective consciousness.

Balance between Individual and Collective

Individual Autonomy

Although collective consciousness is powerful, it should not overshadow individual autonomy. Prosperous societies find a balance between preserving individual rights and freedoms and seeking collective harmony. This delicate balance contributes to social stability and individual flourishing.

Collective Responsibility

Collective consciousness also generates collective responsibility. An individual's actions have repercussions on the entire society, and the responsibility to shape a positive and ethical collective consciousness lies with each member of the community.

Collective Ethics

Collective ethics emerges from how individuals interact within a society. Shared ethical norms, stemming from collective consciousness, define the moral character of a community. These norms guide individual and collective choices, thereby contributing to the construction of an ethically grounded society.

Redefining Consciousness through Quantum Physics

This chapter explores revolutionary ideas and challenges associated with applying quantum concepts to consciousness, while examining future perspectives and the philosophical, ethical, and social implications of this innovative approach.

❖ Quantum Concepts Applied to Consciousness

The fascinating boundary between quantum physics and cognitive processes suggests that quantum principles, such as superposition and entanglement, could play a role in understanding the nature of human consciousness, thereby opening new perspectives on how we perceive and interact with our mental reality.

Fundamental Principles of Quantum Physics

Superposition and Entanglement

Quantum physics is a branch of physics that studies the behavior of subatomic particles at the quantum scale, where the laws of quantum mechanics prevail. Quantum physics relies on astonishing principles that often defy our classical intuition.

Two important phenomena in quantum physics are superposition and entanglement.

- Superposition: Superposition is a fundamental concept in quantum physics. According to the principle of superposition, a quantum particle, such as an electron or a photon, can exist in multiple states simultaneously. This means that until the particle is observed, it can be in a superposition of several different quantum states. It is only when a measurement is made that the particle "chooses" a particular state.

- Entanglement: Quantum entanglement is a phenomenon in which two particles (e.g., two subatomic particles called particles of an EPR pair, or two particles produced during a decay) are linked in such a way that the quantum state of one is instantaneously correlated with the state of the other, regardless of the distance separating them. This means that if the state of one particle is measured, the state of the other particle is immediately determined, even if it is very far away. This phenomenon is often called "spooky action at a distance" and was first formulated by Einstein, Podolsky, and Rosen in their famous EPR (Einstein-Podolsky-Rosen) paradox.

In summary, superposition allows a particle to occupy multiple states simultaneously until it is measured, while entanglement creates a quantum link between two particles so that the state of one is directly related to the state of the other, even at considerable distances.

These strange phenomena of quantum physics have been confirmed by many experiments and have profound implications for our understanding of the fundamental nature of reality at the quantum scale.

Wave-Particle Duality

Wave-particle duality is another astonishing feature of quantum physics, describing the dual behavior of subatomic particles such as electrons and photons. This concept emerged in the early 20th century with the development of quantum mechanics. Wave-particle duality relies on two contradictory aspects of particle behavior at the quantum scale: their corpuscular nature and their wave nature.

- Corpuscular Nature: When particles are observed through the lens of corpuscular nature, they appear to behave like massive particles with position and momentum. This means they can be localized in space and have an identifiable trajectory.

- Wave Nature: On the other hand, when observing the wave behavior of particles, they exhibit wave characteristics. This includes phenomena like diffraction and interference, typical of light waves. Waves do not have a precise position and propagate through space.

The key idea of wave-particle duality is that in the quantum world, particles are neither exclusively particles nor exclusively waves. Their behavior depends on the experimental context and how they are observed.

The iconic experiment illustrating wave-particle duality is the double-slit experiment. When particles such as electrons pass through a double slit, they create an interference pattern, similar to that of light waves. However, when particles are individually measured, they seem to behave like distinct particles, striking a detection screen at specific locations.

The Schrödinger equation, formulated by Austrian physicist Erwin Schrödinger in 1925, is one of the pillars of the mathematical formalism of quantum mechanics, and it plays a central role in describing quantum systems and wave-particle duality.

This duality challenges our intuitive understanding of the behavior of objects on the microscopic scale, but it is fundamental for explaining many observed phenomena in the quantum world.

Indeterminism and Quantum Measurement

Quantum indeterminacy, another fundamental aspect of quantum mechanics, states that certain properties of a particle cannot be precisely predicted before measurement.

Thus, Werner Heisenberg's uncertainty principle, formulated in 1927, states that it is impossible to simultaneously measure with infinite precision the position and momentum (or momentum) of a particle. This means that the more precisely one knows the position of a particle, the less precisely one can know its momentum, and vice versa. This indeterminacy is not due to a technological limitation of measuring instruments but to a fundamental characteristic of the quantum nature of particles.

Quantum indeterminacy also extends to other properties of particles. For example, the spin of particles is subject to quantum indeterminacy. Before measurement, a particle can be in a quantum "superposition" state, where it does not have a defined value for its spin in a particular direction. When a measurement is made, the wave function associated with the particle "collapses" to one of the possible values, and the measurement outcome becomes determined. However, the probabilistic nature of quantum mechanics means that only the probability of obtaining each particular result can be predicted.

Quantum indeterminacy challenges our classical understanding of precisely determining the properties of a particle.

Quantum phenomena are often described in terms of probabilities rather than certainties, and this probabilistic approach is a key characteristic of quantum mechanics.

Analogies Between Human Consciousness and Quantum Phenomena

Principle of Indeterminacy and Free Will

Some philosophers and scientists have suggested an analogy between the principle of quantum indeterminacy and free will in human consciousness. Free will is the concept that individuals have the ability to make autonomous decisions not determined by external factors, involving a form of conscious choice power.

This analogy is based on the idea that, just as quantum particles can be in states of superposition, human decisions can also reside in a state of indeterminacy until a measurement, i.e., a conscious decision-making, is made. In other words, quantum indeterminacy indicates that even with precise initial conditions, certain properties of particles cannot be determined. Similarly, multiple possibilities coexist in human decisions and cannot be determined until a specific choice is made.

This perspective raises the possibility that human consciousness, like quantum reality, has a margin of indeterminacy that transcends deterministic predictions.

Superposition of Thoughts

Quantum superposition has also been envisioned as an analogy with the multiplicity of thoughts and possible mental states in human consciousness.

According to this perspective, one could compare a thought to a quantum particle that can occupy multiple positions simultaneously. Before being observed or "measured" by consciousness, the thought could coexist simultaneously in multiple possible states encompassing various nuances and possibilities until awareness or observation fixes it in a specific state.

This idea is based on the postulate that human consciousness could influence the materialization of a particular thought, much like the act of observation in quantum mechanics fixes the state of a particle. This suggests that our mental experience can be rich in potential and diversity until it is conscious.

Entanglement and Mental Connection

The idea of entanglement implies an instantaneous correlation between distant quantum particles, regardless of the distance between them.

Some have suggested an analogy to a form of mental connection, suggesting that information or mental states could be shared between individuals in a way that transcends spatial boundaries, explaining phenomena such as telepathy or other forms of supposed mental communication.

The underlying idea is that, just as entangled particles can be instantly linked, minds could be connected non-locally, allowing for a form of direct communication without the need for a physical medium.

Emerging Theories on the Quantum Nature of Thought

Orch-OR Theories

A notable theory is proposed by Roger Penrose and Stuart Hameroff, published in the journal "Mathematics and Computers in Simulation" in 1996, called Orchestrated Objective Reduction (Orch-OR).

This theory suggests that consciousness emerges from quantum processes in microtubules, cellular structures present in neurons. The notion of "quantum coincidence" specifically refers to the moment when multiple quantum events occur simultaneously and synchronously in these microtubules, thereby contributing to conscious experience.

Here are the key points of this theory:

- According to quantum mechanics, a particle can exist in a superposition of states, meaning it can occupy multiple states simultaneously until a measurement is made. This superposed state is described by the quantum wave function.

- "Quantum Objective Reduction" suggests that there is a critical point where this superposition of quantum states collapses into a determined state. This collapse is called the reduction of the wave function, and it corresponds to a process by which quantum reality becomes determined.

- Penrose proposes that QOR occurs at the level of microtubules, which are small enough for significant quantum effects to influence their functioning. They are considered key candidates for quantum processes in the brain due to their structure and ubiquity in nerve cells.

- In the context of consciousness, QOR in microtubules is considered a mechanism for quantum choices. This means that the superposition of quantum states in microtubules could represent a multiplicity of possible thoughts or mental states.

The Orch-OR theories have sparked significant criticism in the scientific community. Some scientists argue that the conditions required for macroscopic-scale quantum processes, as proposed by Orch-OR, are challenging to achieve. Moreover, most neuroscientists maintain that consciousness emerges in a more complex and multifactorial way than suggested by the Penrose-Hameroff theory, although the precise details remain unknown.

Nevertheless, research is ongoing to attempt to detect quantum signatures in the brain and validate the propositions of this theory. Here are some elements related to these studies:

- Microtubule Polarization: Studies have examined the polarization of microtubules, which can be influenced by quantum properties. Polarization is an electrical property of microtubules that could play a role in cellular communication. Some researchers have suggested that quantum processes could be involved in this polarization.

- Quantum Coincidence Studies: Research has attempted to detect specific quantum correlations between the quantum states of microtubules. However, these studies are often complex, and their interpretation is controversial.

- In Silico Modeling: Some researchers have used computer models to simulate the behavior of microtubules at the quantum scale. These models aim to explore how quantum phenomena could influence the properties of microtubules and, by extension, neurons.

- Spectroscopy and Microscopy: Advanced spectroscopy and microscopy techniques have been used to study the structure and properties of microtubules at very small scales. This includes approaches such as atomic force microscopy and nuclear magnetic resonance spectroscopy.

- Quantum Decoherence Studies: Quantum decoherence, or the loss of quantum state, is an important aspect to consider in the context of quantum processes in microtubules. Studies have examined how microtubules interact with the environment, which could lead to decoherence and influence the stability of quantum states.

Quantum Information Theory

The perspective that views consciousness as a manifestation of quantum information theory is based on the idea that cognitive processes and the nature of consciousness itself can be understood through principles of quantum mechanics applied to information. Here are some key points of this perspective:

- Quantum Information: In the context of quantum mechanics, information is processed in a particular way. Classical bits are replaced by qubits, which can exist in superposition states, allowing for more complex information processing.

- Superposition and Mental Multiplicity: Some researchers suggest that quantum superposition, which allows a particle to occupy multiple states simultaneously, may be analogous to the multiplicity of thoughts or possible mental states in human consciousness. Before being observed, a thought could coexist in multiple states.

- Quantum Information Integration: Quantum information theory proposes that the integration and manipulation of quantum information in the brain could be related to conscious experience. Some argue that consciousness emerges when a certain quantity or quality of information is reached or processed in the brain.

- Quantum Correlations: Quantum correlations, such as entanglement, could also play a role in consciousness. If particles in the brain were entangled, changes in the state of one particle could instantaneously affect the state of another, which could be related to conscious processes.

Here are some aspects of ongoing research related to this theme:

- Studies on Cognitive Processes: Researchers explore how cognitive processes, such as memory formation and decision-making, could be influenced by principles of quantum mechanics applied to information. Computer models are often used to simulate these processes.

- Studies on Neuronal Networks: Research examines the possibility that neuronal networks in the brain may manifest quantum properties in information processing. This includes studies on quantum coherence and the potential superposition of states in neuronal networks.

- Measurement of Quantum Coherence: Some experiments have sought to measure quantum coherence in biological systems, including components of the brain. However, interpreting these measurements in the context of consciousness remains complex.

- Exploration of Non-local Phenomena: Some work examines the possibility of non-local phenomena in the brain, where events in one part of the brain could instantaneously influence other parts, which could be related to aspects of consciousness.

❖ Challenges in Theoretical and Experimental Research

The attempt to reconcile quantum principles, generally observed at the microscopic scale, with macroscopic phenomena of human thought is complex. This quest raises profound questions about the nature of subjective reality and challenges traditional paradigms of physics and neuroscience, providing a realm of reflection where the boundaries between matter, mind, and perception are questioned.

Conceptual Obstacles Related to Applying Quantum Physics to Consciousness

Macroscopic Scale: Most quantum phenomena are well understood at the microscopic scale, but extending these concepts to the macroscopic scale, where human consciousness operates, is challenging. Decoherence, the process by which quantum systems lose their coherence and become classical, poses a fundamental question about preserving quantum effects in the brain.

The Nature of Observation: The role of observation in quantum physics is debated. How observation by human consciousness could influence quantum processes in the brain raises complex philosophical and conceptual questions.

Non-locality: Quantum non-locality, where particles can be instantly linked over significant distances, may seem incompatible with our understanding of consciousness limited by the speed of neuronal signal transmission.

Sensitivity to Perturbations: Quantum processes are generally considered highly sensitive to environmental perturbations. Some therefore criticize the idea that quantum processes could support complex cognitive phenomena in the human brain, which is constantly exposed to external influences.

Emergence of Consciousness: A major challenge lies in understanding how emergent properties, such as consciousness, could arise from quantum processes. How quantum states and events could translate into a conscious experience remains an enigma.

Current Experimentations Exploring Quantum Aspects of Cognition

Decoherence Studies: Experiments aim to study decoherence in the brain, examining how quantum processes might persist despite environmental influences.

Quantum Sensitivity Measurements: Techniques are being developed to measure quantum sensitivity within neuronal structures, exploring whether specific quantum processes can be detected.

Critiques and Debates Within the Scientific Community

1. **Skepticism:** Some scientists question the idea that consciousness could stem from quantum processes, highlighting the efficacy of classical explanations for many aspects of cognition.

2. **Need for Concrete Evidence:** Most researchers insist on the necessity of strong empirical evidence demonstrating the maintenance of quantum effects at the neuronal scale and their contribution to consciousness.

3. **Divergent Interpretations:** Debates extend to the very interpretations of quantum physics, with different schools of thought regarding the meaning and application of quantum principles in the context of consciousness.

❖ Links Between Consciousness and the Universe

Human consciousness, although often considered individual, could be intrinsically linked to the very foundations of the universe, thus opening intriguing perspectives on the profound nature of existence and our place within the cosmos.

Consciousness as a Fundamental Force of the Universe

The notion that consciousness could be a fundamental force of the universe opens new perspectives on the very nature of existence. Beyond traditional models describing the universe in terms of matter and energy, this idea suggests that consciousness plays an intrinsic role in the cosmic fabric. If matter is the substance from which the universe is woven, consciousness could be the guiding thread that shapes and gives meaning to this fabric. Thus, this perspective proposes that consciousness is not simply a complex emergent product of evolution, but rather a primary force that coexists with the physical laws of the universe. If this is the case, it challenges our very understanding of reality, highlighting that consciousness is not limited to the boundaries of the human mind but is woven into the fabric of the universe itself.

This idea invites a reconsideration of the nature of causality and interconnectivity in the universe. If consciousness is fundamental, then it could be a unifying factor that transcends individual boundaries, connecting all forms of life and existence into a common conscious fabric.

Cosmic Awakening and Conscious Connectivity

Notions of cosmic awakening and conscious connectivity transcend the traditional boundaries of human consciousness.

Cosmic awakening, in this perspective, goes beyond mere individual awareness to encompass an expanded understanding of our place in the universe. It is a spiritual journey that explores the connection between human consciousness and the fundamental principles of cosmic existence. This idea suggests that our consciousness may resonate with the forces shaping the universe, offering a holistic perspective that integrates the individual into the vast fabric of cosmic reality.

Conscious connectivity, in tandem with cosmic awakening, reinforces this idea of unity and interrelation. It involves a profound and interconnected understanding of self, others, and the universe as a whole. This vision transcends the boundaries of the ego to recognize that each individual is an essential element of a larger cosmic fabric. Conscious connectivity emphasizes that the perceived separation between human beings and the universe is an illusion and that individual consciousness may reflect a greater consciousness that permeates all reality.

Potential Links between Individual Consciousness and the Quantum Nature of the Universe

- Panpsychism: Some researchers explore the idea that consciousness could be inherently linked to the very structure of the universe. Panpsychism suggests that consciousness exists at different scales, including the cosmic scale.

- Theories of Unity: Metaphysical perspectives consider the universe as a conscious entity. These theories posit that individual consciousness is intrinsically connected to a universal consciousness, creating a network of interconnection on a cosmic scale.

- Interactions Through the Mind: Some models speculate on mechanisms through which individual consciousness could interact with quantum reality in ways to influence or be influenced by phenomena on a cosmic scale.

Interconnections between Conscious Observation and Quantum Reality

- Role of the Observer: According to the quantum interpretation, the role of the observer is central.

Individual consciousness, as an observer, could play an active role in determining quantum states, raising questions about the co-creation of reality.

- Influence of Expectations and Intentions: Thought experiments suggest that the expectations and intentions of the observer can influence the outcomes of quantum experiments, establishing a link between human consciousness and quantum dynamics.

- Reflections on the Nature of Reality: These interconnections challenge the objective nature of reality, suggesting that quantum reality may be more influenced by conscious perception than previously envisioned.

Recent Developments in Research on Cosmic Quantum Consciousness

- Large-scale Quantum Coherence: Research examines the possibility of large-scale quantum coherence in the universe. This would imply that particular quantum states could be shared across cosmic distances.

- Concepts of Cosmic Consciousness: Some philosophers and scientists explore concepts of cosmic consciousness, suggesting that the universe itself might manifest aspects of consciousness or be intrinsically linked to individual consciousness.

- Spiritual and Metaphysical Perspectives: Discussions in spiritual and metaphysical circles often consider the possibility that human consciousness is connected to a broader reality through quantum links.

❖ Practical Applications and Implications

The practical applications of these often complex metaphysical concepts can have profound ramifications on our daily lives, from technology to medicine. Simultaneously, these advancements raise philosophical implications, prompting fundamental questions about the nature of existence, consciousness, and our relationship to the universe, thus creating a dynamic dialogue between practice and philosophy.

Potential Applications of Quantum Consciousness in Technology

- Conscious Quantum Computing: Some envision quantum computing systems that integrate elements of consciousness. This could lead to quantum computers capable of processing information consciously, or at least in collaboration with conscious processes.

- Quantum Brain-Computer Interfaces: Advances in understanding quantum consciousness could inspire new approaches in the field of brain-computer interfaces. Devices capable of synchronizing with quantum consciousness could enable more sophisticated interaction between the human mind and technology.

- Conscious Algorithms: The development of algorithms inspired by quantum consciousness could lead to more adaptive and intelligent computer systems capable of processing information more similarly to human cognition.

Philosophical and Ontological Changes Resulting from a Quantum Understanding of Consciousness

- Redefinition of Reality: Recognizing quantum consciousness could redefine our understanding of reality itself. The idea that consciousness plays a fundamental role in the nature of the universe could transform our traditional conceptions of objectivity.

- New Perspectives on Identity: The notion that individual consciousness could be intrinsically linked to a universal consciousness challenges classical definitions of individual identity. This could lead to a more holistic vision of existence.

❖ Ethical Debates

The intersection between quantum concepts and the nature of consciousness raises crucial questions about the ethical responsibility of mind manipulation.

Ethical Issues of Conscious Manipulation Through Quantum Paradigms

- Integrity of Conscious Experience: The primary ethical concern lies in respecting the integrity of conscious experience. Any manipulation must ensure that the individual retains control over their consciousness and that it is not altered non-consensually.

- Informed Consent: The issue of informed consent becomes crucial. Individuals must be transparently informed about the implications and potential risks associated with any manipulation of consciousness, and they must have the opportunity to give informed consent.

- Responsibility: Who is responsible for the ethical consequences of consciousness manipulation on a quantum scale? The question of responsibility becomes complex, and appropriate mechanisms must be established to ensure ethical use of these technologies.

Need for Dialogue as Research Progresses

- Research Ethics: Researchers working on quantum consciousness must adhere to rigorous ethical standards. Research protocols must be designed to minimize potential risks to participants, and results must be communicated responsibly.

- Public Dialogue: Open and informed public dialogue is essential for discussing ethical implications. The opinions and perspectives of society must be considered in the development and application of this emerging understanding of consciousness.

- Regulation and Standards: The need for clear regulation and ethical standards in the field of quantum consciousness becomes imperative. Governments, scientific institutions, and civil society must collaborate to establish guidelines that frame the research and use of this knowledge.

❖ Future Perspectives

Research on quantum consciousness paves the way for a deeper understanding of the mysteries of consciousness and could redefine our understanding of reality.

Emerging Trends in Quantum Consciousness Research

- Exploration of Quantum Correlates of Consciousness: Researchers are actively exploring the quantum correlates of consciousness, seeking quantum signatures in mental processes. Thus, experiments aiming to identify quantum phenomena in the brain and consciousness are underway, paving the way for a deeper understanding of the links between quantum physics and thought.

- Unifying Theoretical Models: Attempts to develop unifying theoretical models that integrate consciousness and quantum physics are gaining popularity. These models aim to resolve the apparent duality between quantum reality and subjective experience, proposing conceptual frameworks capable of reconciling these two aspects of existence.

Promising Areas for Future Advancements

- Quantum Technologies for Consciousness Study: The use of advanced quantum technologies, such as quantum computers, offers exciting prospects. These tools could enable more accurate simulations of cognitive processes and the exploration of specific quantum configurations associated with particular states of consciousness.

- Practical Applications of Quantum Knowledge: The fields of medicine, artificial intelligence, and psychology could benefit from the growing understanding of quantum consciousness. Practical applications could include more precise medical treatments, AI algorithms inspired by quantum cognition, and innovative approaches in psychotherapy.

- Study of Altered States of Consciousness: Research on altered states of consciousness, such as deep meditation or psychedelic experiences, could provide crucial insights into the quantum aspects of consciousness. These particular states could serve as privileged observation windows for exploring quantum phenomena in the human mind.

Implications for Understanding the Human Mind and Reality

- Redefinition of the Nature of Consciousness: Advances in research on quantum consciousness could potentially redefine our understanding of the very nature of consciousness. If significant quantum correlates are discovered, this could challenge traditional conceptions of the mind and reality.

- Integration of Consciousness into the Fabric of the Universe: The metaphysical implications of research on quantum consciousness could extend to the idea that human consciousness is intrinsically linked to the fundamental nature of the universe. This perspective could transform how we conceive our place within the cosmos.

- Expansion of the Boundaries of Reality: The discovery of quantum phenomena related to consciousness could broaden the boundaries of reality, questioning the traditional limits between the subjective and the objective. This could open new perspectives on the nature of perception and reality creation.

5

Digital and Spiritual Synergy: Elevating Consciousness through Technological and Spiritual Integration

Spiritual Biohacking: Fusion of Technology and Spirituality

This chapter explores the fascinating intersections between technology and spirituality, examining how spiritual biohacking can elevate consciousness while addressing the challenges that come with this fusion.

❖ Convergence between Technology and Spirituality

Convergence between Technology and Spirituality

Spirituality is a vast and multidimensional concept that encompasses beliefs, values, and practices related to the search for meaning, transcendence, and connection with something greater than oneself. It can be expressed through religious, philosophical, or even secular traditions. Spirituality goes beyond the material aspects of daily life and explores the inner dimensions of existence, often related to existential questions, self-awareness, compassion towards others, and the search for the deeper meaning of life. It may also include mystical, meditative, or contemplative experiences, as well as practices aimed at cultivating inner peace and spiritual well-being. The nature of spirituality varies greatly from person to person and can be approached individually or communally.

Tradition and Innovation

Historically, technology and spirituality have often been perceived as distinct, or even opposing, domains. Technology was generally associated with material progress and efficiency, while spirituality was more closely linked to intangible aspects of existence such as transcendence, connection with the divine, and the quest for meaning.

However, over time, these perspectives have evolved to reflect a more nuanced and integrative understanding. Currently, many people view technology as a potential tool to facilitate and enrich their spiritual lives. From guided meditation apps to online platforms fostering spiritual community, technology offers new opportunities for spiritual exploration and practice. Moreover, the acceleration of technology has generated a sense of spiritual disconnection in some individuals.

In this context, spiritual biohacking emerges as an attempt to reconcile spiritual quest with the tools of modernity. Thus, technology and spirituality are not mutually exclusive but rather complementary, capable of coexisting and mutually reinforcing each other. The emergence of discussions on how technology can be aligned with spiritual values marks a new chapter in the relationship between these two domains, offering unique opportunities for personal growth and global connection.

Scientific Foundations

Although spiritual well-being, often considered within the realm of subjectivity and personal belief, is garnering increasing interest in the scientific community. The scientific basis of spiritual well-being relies on a complex intersection of psychology, neuroscience, sociology, and even genetics. Indeed, studies have shown that regular spiritual practices, such as meditation and prayer, can have measurable positive effects on mental health.

Meditation, for example, has been associated with changes in brain structure, particularly in regions related to attention, stress management, and empathy. Research has also examined the effects of prayer and spirituality on mental health, showing correlations with reduced stress, improved emotional resilience, and increased overall well-being.

At the sociological level, studies have explored the link between participation in spiritual communities and well-being. The social dimension of spirituality, whether manifested in religious gatherings or meditation groups, has been associated with greater life satisfaction and strengthened social support networks.

In terms of genetics, some research suggests that genetic factors may influence individual propensity for spiritual experience. Twin studies have shown that spirituality has a significant genetic component, although interaction with the environment also plays a crucial role.

However, it is important to note that the scientific understanding of spiritual well-being is an evolving field, and the diversity of spiritual experiences makes research complex. While significant advances have been made in understanding the neurobiological and psychological mechanisms, spiritual well-being remains a multifaceted subject that defies monolithic explanation. Nevertheless, these scientific advances pave the way for further exploration of how spirituality and well-being intertwine in the complexity of human experience.

❖ Technologies of Spiritual Biohacking

The concept of spiritual biohacking embodies a synergy between scientific advancements and spiritual practices, seeking to optimize the human experience both on the physical and metaphysical levels.

Approaches such as brain stimulation, the use of consciousness-modulating substances, or even genetic interventions could offer new means to enhance consciousness, stimulate spiritual growth, and push the boundaries of what it means to be human.

Virtual Reality: Immersion in Contemplative Spaces

Virtual reality is an immersive technology that creates a computer-simulated environment, often using special headsets, to provide an interactive sensory experience, giving the user the impression of being physically present in a virtual world.

It offers revolutionary potential in the pursuit of transcendence by allowing individuals to immerse themselves in virtual environments conducive to meditation and contemplation. Thus, virtual reality applications dedicated to relaxation and meditation guide users through visually soothing landscapes, thereby promoting states of inner calm. These virtual environments offer an escape from the daily world, creating spaces where one can connect more deeply with oneself. For example, an individual may find themselves atop a sacred mountain or within an ancient temple, enriching their spiritual quest through virtual exploration of meaning-laden places.

Sound Technologies: Brainwaves and Healing Frequencies

Sound technologies, such as binaural beats and healing frequencies, are increasingly used in the pursuit of transcendence to align the energies of body and mind. These technologies harness the ability of sound to influence states of consciousness.

Brainwaves, generated by the electrochemical activity of the brain, are classified into different frequencies, with the main ones being delta, theta, alpha, beta, and gamma waves. Each frequency is associated with specific states of consciousness, from deep relaxation to attentive wakefulness. These brainwaves play a crucial role in modulating emotions, sleep, and concentration.

- Healing frequencies refer to specific vibrations that promote physical, emotional, and spiritual healing. Some practitioners claim that aligning these frequencies with the body's natural frequencies can stimulate the healing process by restoring energetic harmony. Frequencies such as 432 Hz are particularly emphasized in this context. The underlying idea is that exposure to specific frequencies can positively influence well-being by acting on the nervous system and emotional states.

- Binaural beats, a specific form of sound created by introducing different frequencies into each ear, are gaining increasing interest as a potential means of elevating the human mind. This technique exploits the phenomenon of binaural beating, where the brain perceives a resulting third frequency, inducing specific mental states. Thus, some practitioners assert that binaural beats can promote deep relaxation, improve concentration, and even induce meditative and spiritual states. The underlying idea is that these special sound compositions can influence brainwaves, adjusting frequencies to match particular mindsets. For example, theta and delta frequencies are often associated with deep meditation and sleep, while alpha frequencies are linked to relaxation and creativity.

Guided Meditation Applications: Fusion of Tradition and Technology

Guided meditation applications represent a harmonious fusion of tradition and technology. These applications, available on mobile platforms, offer a variety of meditation sessions tailored to specific goals, from stress reduction to exploring deep consciousness. Virtual guides lead users through meditative practices, thereby facilitating access to transcendent states even for novices. This technology brings an accessible dimension to meditation, removing temporal and geographical barriers. Users can choose meditation sessions that fit their schedules, integrating meditation into their daily routines with increased ease.

Artificial Intelligence and Personalization of Spiritual Experience

Artificial intelligence has also found its place in the pursuit of transcendence by offering increased personalization of the spiritual experience. AI algorithms can analyze individual behaviors, preferences, and emotional responses to recommend specific spiritual practices. For example, an AI-based application could suggest mindfulness exercises based on times of day when the user feels most stressed, or recommend personalized spiritual readings based on topics that interest the individual. This personalized approach enhances user engagement in their spiritual quest by providing recommendations tailored to their specific needs.

Neurofeedback

Traditionally associated with optimizing cognitive performance and managing mental disorders, neurofeedback is gaining increasing interest in the realm of spirituality as an exploratory tool to elevate the human mind and facilitate transcendent experiences.

This method involves measuring brain electrical activity using EEG sensors and then providing visual or auditory feedback, allowing individuals to visualize and regulate their brainwave patterns in real-time. By becoming familiar with this feedback, practitioners can guide individuals toward specific brainwave patterns, promoting particular mental states. In the context of spirituality, this opens the possibility of exploring and cultivating states of mental calm, transcendence, and even deeper spiritual experiences.

Thus, preliminary research suggests that neurofeedback may play a role in facilitating access to deep meditative states. For example, studies have examined the use of neurofeedback to help novice meditators more quickly achieve advanced meditative states, characterized by specific brainwave patterns such as theta and alpha waves. This approach could potentially be used as a catalyst for those seeking to deepen their meditative practice or spiritual exploration.

Individual variability in responses to neurofeedback raises intriguing questions about how this technique can be tailored to personal spiritual paths. Experienced practitioners can customize neurofeedback protocols to meet the specific needs of individuals engaged in various spiritual quests, whether to cultivate mental clarity, promote states of inner peace, or even encourage profound mystical experiences.

Transcranial Stimulation

Transcranial stimulation involves the controlled application of electrical currents or magnetic fields to the brain to modulate its activity. Although these methods were initially used to treat various neurological and psychiatric disorders, researchers are beginning to explore how they could influence the brain processes underlying spiritual experiences.

Thus, transcranial stimulation offers the opportunity to target specific brain regions associated with perception, consciousness, and cognition, opening a window of exploration to understand how these regions might be involved in spiritual experiences. Preliminary studies have examined stimulation of areas such as the dorsolateral prefrontal cortex, which plays a role in decision-making and emotional regulation, and have suggested modifications in self-perception and states of mind more open to spirituality.

❖ Ethics and Spiritual Challenges of Biohacking

Ethical Questions Related to Manipulating Spiritual Experience

Authenticity of Experience

One of the main ethical questions revolves around the authenticity and integrity of spiritual experience. Can technology, by facilitating experiences, alter the nature of spiritual connection? Some argue that technological manipulation compromises the true essence of spirituality.

Informed Consent

Informed consent becomes crucial in the context of spiritual biohacking. Practitioners must be fully informed about the implications of using biohack devices to manipulate their spiritual experience. Questions arise about the transparency of device manufacturers and users' actual understanding of the changes induced by technology.

Risk of Technology Dependency

Endless Quest for Spiritual Experience

The risk of dependency in the quest for transcendence raises major concerns. Can individuals, in seeking technology-facilitated spiritual experiences, become dependent on these devices to achieve altered states of consciousness? This raises questions about spiritual freedom and dependence on external supports.

Potential Effects on Mental Health

Dependency on technology in the pursuit of transcendence can have effects on mental health. Excessive use of biohack devices could potentially lead to issues such as anxiety, depression, or even loss of connection with daily reality. Balancing spiritual exploration and mental health becomes a crucial issue.

Philosophical Challenges Related to the Fusion of Material and Spiritual

Nature of the Soul and the Material

Philosophical challenges in spiritual biohacking extend to the very nature of the soul and the material. How can technology interact with something as intangible and mysterious as the soul? Questions about the duality between the material and the spiritual persist, sparking deep debates about the nature of human existence.

Question of Human Evolution

The fusion of the material and the spiritual raises questions about human evolution. To what extent can technology contribute to the spiritual evolution of humanity? Philosophical challenges address the possibility of a radical transformation of human consciousness.

Ethics of Human Enhancement

Ethical questions related to human enhancement become central in spiritual biohacking. The pursuit of enhanced transcendence raises dilemmas about modifying spiritual experience beyond natural capabilities, questioning the ethics of human enhancement through technology.

❖ Perspectives

Impact on Mental and Emotional Well-being

Biohacking shows significant impact on reducing stress and anxiety. Thus, technological approaches facilitate mental relaxation, inducing states of deep calm that effectively contribute to managing daily stress.

Furthermore, some devices integrate specific programs aimed at improving sleep quality. Guided meditation sessions and neurofeedback techniques are designed to calm the mind, thus promoting restful sleep and physical and mental regeneration.

Moreover, technology-facilitated spiritual experience can play an essential role in strengthening emotional resilience. Regular practices of meditation and spiritual awakening help individuals cultivate a more balanced perspective in the face of life's challenges, thus enhancing their ability to cope with stressful situations. Overall, spiritual biohacking emerges as an integrated approach to enhancing mental well-being, offering targeted technological solutions to alleviate stress, improve sleep, and strengthen emotional resilience.

Emerging Trends

As virtual reality continues to evolve, immersive experiences could allow individuals to experience altered states of consciousness in an even more personalized and meaningful way.

Furthermore, the rapid evolution of brain-computer interface technology could make these devices more accessible to the general public. More user-friendly applications could allow individuals to explore their own mental and spiritual abilities through more sophisticated yet easy-to-use interfaces.

Impact on Society and Belief Systems

The impact of spiritual biohacking on traditional religious practices could be significant. Individuals may choose to explore spiritual experiences in a more individualized manner, challenging the hierarchical structures of established religious institutions.

Moreover, advances in spiritual biohacking could redefine traditional concepts of transcendence. Technology-facilitated altered states of consciousness could broaden the understanding of what it means to transcend the limits of human existence.

The intensive use of biohack technologies could also impact how reality is perceived. The boundaries between the real and the virtual could become blurrier, potentially altering the collective understanding of what is considered "real" in the spiritual context.

Considerations for Responsible Adoption of Technology in the Spiritual Domain

As spiritual biohacking expands, the establishment of ethical standards in the industry becomes crucial. Clear guidelines on transparency, user safety, and environmental responsibility are needed to guide the development of new technologies.

Indeed, spiritual biohacking could be susceptible to abuses and deviations, whether financial, ethical, or even spiritual. Prevention mechanisms for abuse, such as appropriate government regulations and oversight bodies, must be put in place to ensure the integrity of the field.

Additionally, thorough education of practitioners and potential users is necessary. Training programs on ethical aspects, risks, and benefits of biohack technology are essential for prudent use.

Augmented Reality for Elevating the Mind

This chapter explores how augmented reality can be used to elevate the mind, influencing our cognition, creativity, and understanding of reality.

❖ Evolution of Augmented Reality

Definition of Augmented Reality

Augmented reality (AR) relies on overlaying virtual elements, such as images, information, or 3D objects, onto the real world, typically through devices like smart glasses, smartphones, or specific headsets.

The AR technology has undergone significant evolution, transitioning from early rudimentary experiences to more sophisticated applications. Improvements in graphic capabilities, sensors, and display devices have enabled the development of more immersive and accessible AR solutions.

Augmented Reality vs. Virtual Reality

Comparison between AR and VR

AR offers a user experience that combines the real world with overlaid virtual elements, providing augmentation of physical reality. In contrast, VR creates a fully virtual environment in which the user is completely immersed, excluding perception of the real world.

Thus, AR allows for direct interaction with the real environment, providing contextual information in real-time. VR, on the other hand, isolates the user from their physical environment, plunging them into an entirely new and pre-fabricated reality.

Distinct Advantages of AR

AR seamlessly integrates into daily life, offering relevant information without disrupting interaction with the real world. This allows for continuous use of the technology without total disconnection from the environment.

Moreover, AR does not require heavy equipment, as is often the case with VR utilizing immersive headsets. AR applications can be used on wearable devices such as smartphones and smart glasses, offering more convenient accessibility.

Possibilities of Convergence of Both Technologies

The convergence of both technologies gives rise to mixed reality, which combines elements of the real and virtual worlds in an interactive manner. This creates even richer experiences, offering both the immersion of VR and seamless integration of AR.

Thus, AR and VR can be used complementarily to create more holistic experiences. For instance, AR can provide contextual information in a museum, while VR can transport the user to a particular historical era.

Current Applications of AR in Various Fields

AR in Education

AR transforms learning by introducing interactive elements into textbooks, providing real-time simulations, and creating more engaging learning experiences. Educational applications also allow students to explore complex subjects visually.

AR in the Medical Sector

In medicine, AR is used for surgeon training through simulations, 3D visualization of organs for surgical planning, and even guided interventions. This allows for increased precision and better understanding of anatomical structures.

Industrial Applications of AR

In the industry, AR is deployed for predictive maintenance, worker training, and visualization of 3D models in real-world contexts. It enhances operational efficiency by providing contextual information to workers in the field.

AR in Entertainment

AR has revolutionized entertainment by offering immersive games, augmented virtual reality experiences, and interactive applications in the artistic sector. AR filters on social media are a popular example.

❖ Impact on Perception and Cognition

AR and Reality Perception

Integration of Virtual and Real

AR transcends the boundaries between the virtual and real world, creating a unique fusion of these two realities. This overlay of digital elements on the real world fundamentally alters our perception, allowing us to see and interact with digital information in our physical environment.

Enrichment of the Physical Environment

A distinctive feature of AR is its ability to enrich our physical environment. Contextual information, visual annotations, and virtual objects can be added to our daily reality, providing a new layer of information and thus transforming our perception of space and objects.

Customization of Sensory Experience

AR allows for extensive customization of the sensory experience. AR filters applied to real objects, contextual information projected onto images, or virtual elements integrated into real environments can be tailored to individual preferences, thus offering a personalized reality to each user.

Cognition, Attention, and Memory

Impact on Cognition

AR has a significant impact on cognition by actively engaging users in interactive experiences. The ability to interact with virtual elements promotes active learning and can even enhance understanding of complex concepts through immersive visual representations.

Modulation of Attention

The integration of virtual elements into our field of vision changes how we allocate our attention. Users can choose to focus on specific virtual information, which has implications in areas such as training, user interface design, and information management.

Effects on Memory

AR also has effects on memory. Visual information associated with virtual elements may be better memorized, and the overlay of virtual reminders in real environments can enhance spatial memory. However, it is important to note that cognitive overload can have negative effects on memory in some cases.

Developing Problem-Solving Skills

AR scenarios can be designed to simulate complex problems, thus encouraging users to develop their problem-solving skills. Interactive puzzles, simulations of professional situations, or educational games can be leveraged to foster an analytical and creative approach to challenges.

Creativity and Critical Thinking

Virtual Creation in the Real World

AR allows for virtual creation directly in the real world, providing fertile ground for creativity enhancement. Three-dimensional drawing applications, virtual sculptures, or even the overlay of digital artworks on physical surfaces encourage creative expression.

Simulation of Complex Scenarios

Critical thinking can be enhanced through the simulation of complex scenarios. Users can be immersed in virtual situations that require thoughtful and strategic decision-making. This can be particularly useful in areas such as professional training, crisis management, or education.

Measurement and Tracking of Improvements

AR applications can integrate tools for measuring and tracking mental performance. Users can receive real-time feedback on their performance in various cognitive tasks, facilitating an iterative approach to personal improvement.

The inherent flexibility of AR allows for adaptability of exercises based on the user's skill level. Challenges can be dynamically adjusted to maintain an optimal level of cognitive stimulation, thereby avoiding boredom or frustration.

Simulating Altered States of Consciousness

Immersion and Transcendence

AR offers exceptional potential for simulating altered states of consciousness through immersion in virtual environments. When users are fully engaged in AR experiences, they may experience a transcendence of ordinary reality, diving into captivating virtual worlds that defy the limits of traditional perception.

Expanded Sensory Experimentation

AR applications can expand our sensory experimentation by introducing visual, auditory, and even haptic stimuli in virtual contexts. This can lead to unique sensory experiences and exploration of reality that goes beyond conventional boundaries.

Exploration of Consciousness and Reality

AR can also be used as a tool for exploring consciousness, allowing users to experience alternative perspectives and constructed realities. This raises philosophical questions about the nature of reality and consciousness, paving the way for deep reflections on our understanding of the world.

❖ Challenges of Augmented Reality

Technical Obstacles and Current Limitations

Hardware and Cost

The quality of the AR experience largely depends on the hardware used. Currently, advanced AR devices can be expensive, thus limiting access to these technologies for many people. Progress in cost reduction and hardware compatibility is essential for broader adoption.

Technological Constraints

Current AR may be limited by technological constraints such as processing power, battery life, and connectivity. These limitations can affect the fluidity and quality of the user experience.

Potential Impacts on Mental Health

Risk of Dependency

Excessive use of AR can potentially lead to dependency issues. Constant immersion in virtual environments can result in a disconnection from physical reality, negatively impacting mental health by creating behavioral addiction.

Effects on Reality Perception

Overlaying virtual elements on physical reality can alter the user's perception, creating a hybrid reality. This raises questions about how these alterations may influence mental health, especially regarding the distinction between the real and the virtual.

Ethical Challenges

Privacy Concerns

The widespread integration of AR raises ethical concerns related to privacy. It is essential to establish robust policies to protect users' personal information and prevent any form of unauthorized surveillance.

Equitable and Inclusive Access

The development of AR should aim to ensure equitable and inclusive access to these technologies. Socio-economic considerations must be taken into account to avoid creating disparities in access and use of AR.

Security and Abuse Prevention

User safety should be a priority, with measures in place to prevent potential abuses of the technology. Developers should collaborate with security experts to minimize the risks of hacking and manipulation.

❖ Future Perspectives

Evolution of AR Devices

Future AR devices should be lighter, more compact, and offer an enhanced user experience. The integration of more advanced sensors, such as biometric sensors for measuring stress and brain activity, could enable more sophisticated applications.

Integration into Daily Life

AR should become an integral part of daily life, with applications ranging from enhanced navigation to personalized contextual information delivery. Wearable devices, such as smart glasses, could become commonplace accessories, allowing continuous use of AR.

Developments in Education

In the educational field, AR could revolutionize learning by providing immersive experiences. Virtual classrooms, real-time historical site visits, and interactive simulations could transform how students acquire knowledge.

New Forms of Entertainment

AR is also expected to revolutionize the entertainment industry by offering more immersive and interactive experiences. Games, movies, and artistic experiences could leverage AR to create engaging virtual worlds.

Applications for Mental Health

AR could play a crucial role in promoting mental health by offering relaxation, meditation, and stress management experiences. Specific applications for treating phobias and anxiety disorders could also be developed.

Major innovations are expected in the medical field, with AR applications used for surgical training, neurological rehabilitation, and pain management. AR could also be integrated into early diagnosis of neurological disorders.

Personal Development

AR can be a powerful tool for personal development. Applications focused on meditation, mindfulness, and stress management can be created to promote mental and emotional well-being.

AR could also offer virtual assistance to help individuals achieve their personal goals. Virtual coaches could guide users in their professional, educational, and personal development.

6

Beyond Mental Boundaries: Advances in Brain-Machine Connection

Technological Telepathy: When Thought Becomes Universal Language

This chapter explores the rapid evolution of brain-computer communication, the possibilities of a universal language of thought, as well as the practical applications and ethical challenges associated with this form of technological telepathy.

❖ Brain-Computer Communication

History of Brain-Computer Communication

Early Experiments

The beginnings of brain-computer communication (BCC) trace back to the 1920s when electroencephalography (EEG) was developed to record brain electrical activity. However, the early experiments of BCC were conducted in the 1970s. Researchers used EEG to enable subjects to generate electrical signals by thinking about specific movements, thus laying the groundwork for direct communication between the brain and computer.

Emergence of Brain-Computer Interfaces

The 1990s marked a crucial milestone with the emergence of brain-computer interfaces. Researchers developed systems allowing individuals to control computer cursors using only their thoughts. This decade also witnessed the introduction of advanced brain imaging techniques, expanding the possibilities of BCC.

Revolution of Neural Interfaces

Over the past two decades, BCC has undergone a revolution with the development of more sophisticated neural interfaces. Electrodes implanted directly into the brain, flexible electrodes, and advanced imaging technologies have opened new perspectives, enabling faster and more precise communication between the brain and computers.

Technologies in Brain-Computer Communication

Electroencephalography

EEG is one of the oldest technologies in BCC. It records brain waves generated by neuronal activity on the surface of the skull. Although non-invasive, EEG offers limited spatial resolution and is primarily used for applications such as game control and mental state detection.

Brain-Computer Interfaces (BCI)

BCIs are systems that enable direct communication between the human brain and a computer or other electronic device. These interfaces aim to interpret the electrical signals generated by the brain and translate them into commands to control external devices. Here is a general explanation of their operation:

- Capturing brain signals: BCIs typically use EEG, functional magnetic resonance imaging (fMRI), or other methods to measure brain activity. EEG is one of the most commonly used methods.

- Data processing: Brain signals captured by EEG are then processed by computer algorithms. These algorithms analyze patterns of brain activity to extract useful information, such as movement intentions, mental concentration, or other cognitive states.

- Interpreting intentions: Once the data is interpreted, the brain-computer interface translates this information into commands understandable by an external device. For example, if a person thinks about moving their right arm, the interface could generate a command that moves a cursor on a computer screen to the right.

- Controlling an external device: The commands generated by the brain-computer interface are then used to control external devices such as computers, robotic prostheses, electric wheelchairs, or other applications.

Implantable Neural Interfaces

Implantable neural interfaces are medical devices that are implanted directly into an individual's nervous system. These interfaces are designed to enable bidirectional communication between the brain (or other parts of the nervous system) and external devices such as computers, prostheses, or other technologies. Here are some key points about implantable neural interfaces:

- Surgical implantation: Unlike non-invasive BCIs, implantable neural interfaces generally require surgical intervention to be implanted. This may involve attaching electrodes directly to the surface of the brain (cortical electrodes) or inserting them into brain tissue.

- Capturing neuronal signals: Implanted electrodes capture the electrical signals produced by neurons. These signals can be used to read neuronal activity and understand movement intentions, thoughts, or other information generated by the brain.

- Electrical stimulation: In addition to signal capture, some implantable neural interfaces also allow electrical stimulation of neurons. This can be used to modulate brain activity for the purpose of treating medical conditions or enhancing cognitive performance.

- Medical applications: Implantable neural interfaces are often used for medical purposes. For example, they can be used to treat neurological disorders such as epilepsy, Parkinson's disease, or treatment-resistant depression. They are also being studied to restore motor functions in paralyzed individuals.

Recent Progress and Implications

Advancements in Precision and Speed

Recent progress in BCC focuses on improving the precision and speed of communication. Machine learning algorithms have been integrated to allow systems to adapt to individual brain activity patterns, thereby enhancing the reliability of commands.

Medical and Rehabilitation Applications

Medical applications are significant. Devices are being developed to assist individuals with paralysis, neurological disorders, or brain injuries in regaining some autonomy. Additionally, these techniques can be used in rehabilitation to help restore motor function in individuals with physical impairments.

Advanced Human-Machine Interactions

Advancements in BCC pave the way for more advanced human-machine interactions. Systems enable the control of physical objects, manipulation of virtual environments, and even transmission of information directly from the brain to other devices, thereby expanding the possibilities of human interaction with technology.

Support for Mental Illnesses

BCC is also explored as a potential tool in managing mental illnesses such as depression and anxiety. By monitoring brain activity patterns, the technology could help identify early signs of mental disorders and provide tailored interventions.

❖ Universal Language and Mental Connection

Exploration of the Concept of Universal Language of Thought

Definition of Universal Language

The concept of a universal language of thought suggests the existence of a means of communication that transcends linguistic and cultural barriers, enabling direct understanding between individuals at the level of thought. It seeks to address the fundamental question of whether thoughts can be shared in a way that goes beyond the limits of spoken languages.

Universal Non-Verbal Communication

Researchers have long explored the idea of universal non-verbal communication, where certain facial expressions, gestures, or postures are considered understandable regardless of culture. However, the universal language of thought goes beyond, seeking to establish a direct connection at the level of ideas and concepts.

Technological Implementation of Universal Language

In the technological realm, the idea of a universal language of thought has led to research on BCIs capable of decoding and transmitting thoughts directly. These interfaces aim to create a form of communication that does not require linguistic translation but rather transmits the essence of thought.

Neural Bases of Mental Communication

Mirror Neurons and Empathy

Mirror neurons, discovered in the 1990s, play a crucial role in understanding mental communication. They are activated when we observe or imagine an action, creating neural resonance between the observer and the actor. This activation of mirror neurons is linked to empathy, suggesting a neural basis for the intuitive understanding of others' mental states.

Brain Networks Involved in Abstract Thought

Abstract thoughts, such as ideas and concepts, are related to the activation of specific brain networks. Associative cortical areas, such as the prefrontal cortex, are involved in the representation and manipulation of more complex concepts. Understanding these neural mechanisms paves the way for the possibility of translating these complex thoughts into understandable signals.

Electroencephalographic Activity

Brain imaging techniques, such as EEG, allow real-time observation of brain electrical activity. Studies using EEG have shown that specific patterns of brain activity are associated with particular mental states, thus providing clues for translating thoughts into language.

Translation of Thoughts into Understandable Language

Decoding Brain Patterns

Translating thoughts into understandable language relies on the ability to decode patterns of brain activity associated with specific concepts. Machine learning algorithms and artificial intelligence play a crucial role in this task, allowing for the association of brain activity patterns with specific meanings.

BCIs for Mental Communication

BCIs represent one of the technological avenues for translating thoughts into understandable language. By using sensors such as EEG, these interfaces can detect brain activity patterns and associate them with specific commands or intentions, enabling direct thought-based communication.

High-Resolution Brain Imaging

The evolution of brain imaging technologies, including high-resolution fMRI, offers the opportunity to observe finer details of brain activity patterns. This could enable more precise translation of thoughts, going beyond simple commands to include ideas and more complex concepts.

❖ Practical Applications

Medical Applications

Communication Restoration

BCC offers significant hope for paralyzed individuals, particularly those with spinal cord injuries or neurodegenerative diseases. Using BCIs, these individuals can regain a form of communication by translating their thoughts directly into commands for computers.

Control of Prosthetics and Exoskeletons

Beyond communication, BCC also enables control of prosthetics and exoskeletons. Paralyzed individuals can use their thoughts to direct the movements of artificial limbs, thereby offering greater autonomy and improved quality of life.

Improvement of Quality of Life

The impact of BCC on the quality of life of paralyzed individuals extends beyond communication and mobility. These technologies can also help reduce dependence, allowing individuals to perform daily tasks autonomously, which has profound implications for mental and emotional health.

Acceleration of Medical Research

BCC is also a valuable tool for medical research. By allowing researchers to interact directly with the brains of study subjects, BCC can accelerate the understanding of underlying mechanisms of certain medical conditions, paving the way for new treatments and therapies.

Other Applications

Military Applications

In the military and security domain, BCI can be used for remote control of equipment. Soldiers could use their thoughts to interact with drones, vehicles, and other devices on the battlefield, thereby providing increased operational flexibility.

BCI can also be used to enhance cognitive performance of operators. By enabling direct communication between the brain and computer systems, decisions and actions can be executed more rapidly, which is crucial in military environments where reaction speed can make a difference.

Applications in Education and Learning

For students with special needs, such as those with paralysis or communication disorders, BCI can be a valuable resource. It offers an alternative pathway to express ideas, participate in discussions, and access educational materials.

BCI can also allow for more personalized learning. BCIs can detect brain activity patterns associated with attention, comprehension, and engagement, automatically adjusting educational content based on each student's needs.

BCI also paves the way for the development of new teaching methods. Teachers could use these technologies to create immersive learning experiences based on students' brain activity, thereby fostering a deeper understanding of concepts.

Limits and Technological Challenges

BCI presents several technological challenges that must be overcome to make these interfaces more accurate, reliable, and accessible. Here are some of the most important challenges faced by researchers and engineers working in this field:

Brain Complexity

Achieving seamless communication via BCI is hindered by the intrinsic complexity of the human brain. The brain is a remarkably complex organ with billions of interconnected neurons, forming dynamic networks that are constantly evolving. Precisely understanding these networks and their activities is a monumental challenge.

Interindividual Variability

Each brain is unique, exhibiting significant interindividual variability. Brain activity patterns can vary considerably from one person to another, making it difficult to establish universal models for BCI. Algorithms must be sufficiently flexible to adapt to individual differences, adding a layer of complexity to the development of these technologies.

Dynamics of Thoughts and Emotions

Thoughts and emotions are dynamic and changing processes. The ability of BCI to track these fluctuations in real-time is a major challenge. Current interfaces may struggle to capture the richness of human thoughts, thereby limiting the fluidity of communication.

Spatial and Temporal Resolution

Current technologies have limitations in terms of spatial and temporal resolution. The ability to detect and interpret complex thoughts, especially those related to abstract concepts, remains a major challenge.

Precision and Reliability

Brain electrical signals can be weak and susceptible to interference, making their interpretation complex. Advances in sensor quality and data processing algorithms are needed to improve accuracy.

Selectivity

BCIs must be able to specifically distinguish signals related to the user's intention from the electrical noise generated by other brain activities. This is particularly important for applications such as prosthetic control, where precision is crucial.

Long-Term Stability

Some devices may lose effectiveness over time due to the immune system's reaction to the presence of electrodes or other implantable components. Research is focused on developing biocompatible materials and strategies to maintain long-term stability of these devices.

Non-Invasive Interfaces

While invasive interfaces often offer better signal resolution, their use requires surgical intervention. Non-invasive brain-computer interfaces, such as those based on EEG, are easier to deploy but generally offer lower resolution. Improving the performance of non-invasive interfaces is a significant challenge.

Integration with Prosthetics and External Devices

The interface must be able to provide effective commands to control prosthetics, electric wheelchairs, computers, etc. This requires close integration between the interface and the external device, as well as adaptive control algorithms.

❖ Ethics of Technological Telepathy

The idea of reading thoughts raises fundamental questions about the right to private thought. In many societies, thinking is considered one of the last intimate and personal refuges. Technological access to thoughts raises the question of whether this mental sanctuary can be preserved and respected.

Consent

Reading thoughts by BCI devices raises crucial questions of consent. If a person has not explicitly consented to sharing their thoughts, collecting and interpreting those thoughts by third parties may be considered a privacy violation. This underscores the need to establish clear consent standards.

Accuracy and Correct Interpretation

Another ethical challenge is related to the accuracy of thought reading and the correct interpretation of those thoughts. Current technologies may not be perfectly accurate, which could lead to misinterpretations. Errors in thought reading could have serious consequences, including false accusations or erroneous judgments.

Legal Protection of Cognitive Rights

Cognitive rights, including the right to mental privacy and cognitive autonomy, should be protected by appropriate legal frameworks. These legal protections must evolve to reflect technological advances and ensure that individuals' fundamental rights are preserved in an increasingly mentally connected world.

Manipulation of Thoughts and Perception

The ability of technological telepathy to influence thoughts raises significant ethical concerns. Manipulation of thoughts and perception can be used for positive purposes, but it can also be exploited maliciously. The need to prevent any form of unconsented external control becomes imperative.

❖ Future Perspectives

Future Developments of Technological Telepathy

Advanced Brain-Computer Interfaces

Future developments in technological telepathy are likely to be marked by significantly improved BCIs. Progress in spatial and temporal resolution will enable more precise reading of thoughts, paving the way for smoother and more detailed communication.

Advanced Emotion Interpretation

The evolution of technological telepathy will likely include an increased ability to interpret emotions. Beyond the transmission of ideas, future technologies may be able to transmit and receive complex emotions, enriching communication with a deeper emotional dimension.

Enhanced Bidirectional Communication

Future developments could also enable enhanced bidirectional communication. Rather than simply reading thoughts, technological telepathy could allow for interactive exchanges where both parties can contribute simultaneously to the mental conversation, creating a more dynamic communication.

Integration with Other Technologies

Technological telepathy could be integrated with other technological advances. Integration with augmented reality, virtual reality, or other immersive interfaces could expand communication possibilities, allowing users to share virtual experiences more immersively.

Potential Impacts on Society and Interpersonal Communication

Reduction of Linguistic Barriers

Technological telepathy has the potential to reduce linguistic barriers. If thoughts can be transmitted directly, the need for linguistic translation could decrease, facilitating communication between people speaking different languages.

Enhancement of Empathy and Understanding

The impact on interpersonal communication could be profound by enhancing empathy and understanding. By allowing individuals to share their thoughts and emotions directly, technological telepathy could create closer bonds and better mutual understanding.

New Forms of Social Interaction

Future developments of technological telepathy could give rise to new forms of social interaction. Virtual spaces where individuals can interact mentally could emerge, creating communities based on brain connectivity rather than geographical proximity.

Anticipated Evolution of Language and Human Understanding

Changes in Language Perception

The introduction of technological telepathy could lead to profound changes in language perception. Verbal expression could lose its relative importance compared to mental communication, thus altering how individuals attribute meaning and understand ideas.

Adaptation of Cognitive Processes

Adapting to new technologies could also influence human cognitive processes. Technological telepathy could stimulate changes in how individuals think and process information, thus reshaping human cognition in unexplored ways.

New Forms of Creative Expression

The evolution of language and human understanding could give rise to new forms of creative expression. The ability to communicate mentally could inspire new forms of art, storytelling, and expression that transcend the boundaries of traditional media.

Downloading Consciousness and Brain Immortality

This chapter delves into technological advancements, philosophical implications, and ethical challenges associated with the attempt to download human consciousness.

❖ Possibilities of Consciousness Transfer

Concept of Consciousness Uploading

Consciousness upload refers to the notion of transferring all the information and mental processes of an individual from a biological substrate, such as the brain, to a non-biological medium, like a computer.

This goes beyond simple artificial intelligence simulation, aiming to capture the very essence of individual consciousness. This process would involve converting all of a person's experiences, thoughts, emotions, and memories into digital data, thereby enabling the creation of a digital replica of the human mind.

At the core of this notion lies the idea that consciousness, instead of being intrinsically tied to the biology of the brain, could be encapsulated and transferred to an artificial substrate, such as a sophisticated computer network.

This approach raises profound questions about the nature of identity, autonomy, and personal persistence. Researchers are exploring the fields of neuroscience, computer science, and artificial intelligence to understand the complexities of consciousness and the technological possibilities that could eventually enable such a transfer.

However, the ethical and philosophical implications of this idea spark intense debate, challenging our traditional conceptions of human existence, mortality, and individual singularity.

Scientific and Technological Approaches

Neural Connection Mapping

One approach involves meticulously mapping all the connections between neurons in the human brain, a discipline known as connectomics. Projects such as the Human Connectome Project seek to understand brain connectivity using advanced brain imaging techniques.

In this context, Diffusion Magnetic Resonance Imaging (dMRI) allows visualization of nerve fiber pathways, and at a finer scale, Scanning Electron Microscopy (SEM) enables observation of ultrastructural details of synaptic connections between neurons. This contributes to a deeper understanding of how signals are transmitted between brain cells. Additionally, EEG and intracranial EEG (iEEG) can record brain electrical activity, providing insights into neuronal communication patterns.

Based on the collected data, detailed computer simulations can be conducted to model brain behavior, cognitive processes, and emotional processes.

Progressive Mind Transfer

Some scenarios suggest that rather than instantly transferring a person's consciousness to a digital medium, it could be a gradual process, where the mind is transferred progressively, perhaps by gradually replacing certain parts of the brain with artificial components.

Current BCIs already allow for some form of interaction between the human brain and external devices. Thus, advancements in this field could enable a gradual transfer of certain cognitive or motor capabilities to computer systems.

Indeed, machine learning technologies can be used to analyze and mimic human cognitive behaviors. A progressive transfer could begin by delegating some mental tasks to AI systems, gradually expanding the field of transferred skills.

Moreover, a progressive model could focus on selectively uploading memories. Devices could be used to record and store life experiences, thus creating a form of personal digital archive.

Neural Implants

The idea of using neural implants for consciousness uploading involves integrating implantable devices into the brain to facilitate information transmission between the human brain and a digital system.

Here are some aspects to consider:

- Recording Information: Neural implants can record neuronal activity in real-time, which could provide crucial data on brain function.
- Information Storage: Neural implants could record specific patterns of brain activity related to thoughts, memories, or emotions and could potentially be used to create backup copies or facilitate selective uploading of certain mental information.

❖ Current Scientific Realities

Scientific and Technological Limitations

Complexity of the Brain

The complexity of the human brain, with its billions of interconnected neurons, poses significant challenges. Understanding how these connections give rise to consciousness is a major challenge. Reproducing consciousness requires a significant advancement in our understanding of cognition, memory, and complex neuronal processes.

Subjective Nature of Experience

Conscious experience is deeply subjective, and capturing all its dimensions through objective means remains difficult. Faithfully reproducing this subjectivity remains a major scientific and technological barrier.

BCI Limitations

Current BCIs are primarily focused on medical applications and prosthetics, but the complex bidirectional communication required for consciousness uploading is well beyond their current capabilities.

Quantum Computing

While quantum computing offers intriguing prospects in terms of computational power, it is still in the development stage, and its application to consciousness uploading raises unresolved questions.

Future Perspectives

Interdisciplinary Research

The field of consciousness uploading will require close collaboration between neuroscience, computer science, philosophy, and other disciplines to make significant progress.

Development of New Technologies

Advancements in neurotechnologies, quantum computing, and other technological domains are expected to play a crucial role in future research.

More Modest Approaches

More modest approaches, such as digitizing memories or developing more sophisticated BCIs, could be initial steps before considering full consciousness uploading.

❖ Elimination of Bodily Needs

Immortality of the Body

The idea of achieving bodily immortality is an intriguing concept. Here are some aspects related to this concept:

Artificial Organs and Transplantations

- Regeneration of Organs: Advances in bioengineering could enable the creation of artificial organs capable of regeneration, thereby prolonging lifespan by replacing failing organs.

- Advanced Transplantations: More sophisticated transplantation techniques, including the use of artificial tissues and genetic modification techniques, could improve compatibility and lifespan of transplanted organs.

Gene Therapies and Cellular Regeneration

- Cellular Regeneration: Gene therapies can help regenerate cells and repair genetic damage, offering potential to slow the aging process.
- Enhancement of Regeneration Capabilities: Research could focus on enhancing natural cellular regeneration mechanisms to maintain tissue and organ health.

Artificial Intelligence and Health Monitoring

- Early Diagnosis: Using artificial intelligence allows continuous health monitoring, identifying potential issues as they arise and enabling early intervention.
- Treatment Adaptation: Integrated systems could adjust medical treatments in real-time based on changing body needs.

Regenerative Medicine and Nanotechnology

- Molecular-scale Repair: Advances in regenerative medicine and nanotechnology could enable cellular and molecular repairs, thereby extending lifespan.
- Elimination of Defective Cells: Technologies to target and eliminate defective or damaged cells could slow the aging process.

Advancements in Human-Machine Integration

In recent years, advancements in human-machine integration (HMI) have been significant, opening new perspectives in various fields, including neuroscience, computer science, robotics, and health.

Here are some specific areas where notable progress has been made:

Brain-Computer Interfaces

Prosthetic Control: Advanced BCIs allow individuals to control prosthetic limbs more precisely using brain signals.

Neurological Rehabilitation: BCIs are increasingly used in rehabilitation after brain injuries, facilitating relearning of movements and motor skills.

Quantum Computing

More Powerful Calculations: Quantum computers promise significantly improved computational capabilities, which can have a significant impact on brain modeling and simulation.

Data Security: Quantum technologies also offer opportunities to enhance data security, which is crucial for HMI applications related to the confidentiality of brain information.

Robotics and Intelligent Prosthetics

Sensor Integration: Smart prosthetics and robots increasingly integrate advanced sensors for better environment perception and more natural interaction with users.

Machine Learning: Robots with machine learning capabilities can adapt to individual preferences and needs, thereby enhancing their usefulness in various contexts.

Virtual and Augmented Reality

Immersive Interfaces: Virtual and augmented reality technologies enable immersive experiences that can be used for training, therapy, and other HMI applications.

Improved Communication: Integrating augmented reality into user interfaces can improve communication by overlaying useful information in the field of vision.

Connected Health and Telemedicine

Continuous Monitoring: Connected health devices enable continuous monitoring of physiological parameters, improving the management of chronic diseases and the prevention of health issues.

Early Interventions: HMI systems in the healthcare domain facilitate early interventions by providing real-time alerts and improving access to healthcare.

Artificial Intelligence

Brain Data Processing: AI techniques are increasingly used to analyze and interpret brain data, facilitating the understanding of complex patterns.

Decision Support: AI plays a growing role in decision support systems, enhancing the efficiency of human-machine interfaces in areas such as navigation, energy management, and design.

Natural Interfaces and Voice Commands

Improved Recognition: Voice-based and gesture-based interfaces are becoming more accurate, offering more natural interaction with devices.

Brain Commands: Although still in development, research on brain commands is progressing, exploring the possibility of controlling devices through thought.

Symbiosis and Enhancement of Human Abilities

HMI aims not only to compensate for deficiencies but also to augment human capabilities beyond their natural limits. Thus, exoskeletons and cognitive augmentations seek to amplify physical strength and intellectual abilities, creating synergy between humans and machines.

Exoskeletons

Enhancing human capabilities through exoskeletons represents a significant advancement in the fields of robotics and assistive technologies. These body-worn devices are designed to enhance human physical abilities, whether in terms of strength, endurance, or mobility.

Here are some key aspects of enhancing human capabilities with exoskeletons:

- Physical Strength Enhancement: Exoskeletons are designed to increase human muscle strength, allowing users to lift and manipulate heavier loads without straining their muscles.

- Mobility Assistance: Exoskeletons can assist people with mobility limitations by providing support and facilitating walking. This is particularly beneficial for individuals with neurological disorders or spinal cord injuries.

- Rehabilitation and Physical Therapy: Exoskeletons are used in rehabilitation programs to help patients recover from injuries or surgeries by providing physical support and facilitating repetitive movements.

- Military and Industrial Applications: In the military domain, exoskeletons are explored to enhance soldiers' strength and endurance. In industry, they can be used to improve productivity by reducing worker fatigue.

- Brain-Exoskeleton Interfaces: Some research explores brain-exoskeleton interfaces, where brain signals are used to control the exoskeleton's movements, or vice versa.

- Medical Exoskeletons: Medical exoskeletons are developed to assist people with paralysis in regaining some mobility. These devices are designed to be worn permanently and can be controlled by specific body movements.

- Injury Risk Reduction: In industrial or construction environments, exoskeletons can reduce the risk of injuries by relieving physical stress on the human body.

- Adaptation to Extreme Conditions: Some exoskeletons are designed for use in extreme conditions, such as underwater environments or emergency situations, allowing users to maintain their effectiveness in challenging circumstances.

While challenges persist, such as component miniaturization, battery autonomy, and natural adaptation of the human-machine interface, with continuous progress in technology, these devices have the potential to transform many aspects of daily life, from healthcare to professional and military domains.

Cognitive Augmentations

The symbiosis between humans and machines also opens new frontiers in cognitive enhancement. Here are some approaches being explored to amplify intellectual capabilities through this fusion:

- Concentration Enhancement: Brain-computer interfaces (BCIs) can be used to improve concentration and mental focus by providing instant feedback on brain activity. This can be useful for tasks requiring sustained attention.

- Thought-Controlled Devices: The ability to control electronic devices or prosthetics through thought can extend physical and functional capabilities.

- Mnemonic Implants: Neural implants can be explored to enhance memory by facilitating more efficient recording and retrieval of information.

- Direct Learning Interfaces: Interfaces that allow direct learning through the brain, by downloading specific knowledge or skills, could accelerate the process of acquiring new abilities.

- Augmented Calculations: Integrating augmented calculation tools, where the human brain is connected to advanced computer systems, can speed up information processing and solving complex problems.

- Access to Extended Databases: Human-machine fusion can enable direct access to extended databases, providing instant information for decision-making.
- Direct Brain-to-Brain Communication Interfaces: Interfaces allowing direct communication between brains could revolutionize how ideas and concepts are shared, enabling faster and more precise communication.
- Instantaneous Thought Translation: The ability to instantaneously translate thoughts from one language to another could facilitate communication in an increasingly connected world.
- Real-Time Assistance: Thought-integrated assistance systems could help individuals solve problems or accomplish complex tasks.
- Creative Brain Stimulation: Devices can be developed to specifically stimulate brain centers related to creativity, facilitating the generation of innovative ideas.
- Monitoring and Early Intervention: Integrated systems can continuously monitor signs of stress or mental health issues, offering preventive interventions.
- Emotional Well-Being Enhancement: Devices can be designed to regulate emotional states, promoting resilience and well-being.

Thus, human-machine fusion could be the first step towards consciousness uploading. Advanced brain-machine interfaces could enable direct bidirectional communication between the human brain and a computer system. These interfaces could record information about brain activity patterns, memory, emotions, and so on.

Over time, as these technologies evolve, a progressive transition process could be imagined. Specific neural data could be extracted and transferred to digital substrates, while maintaining a connection with the biological brain. This could allow for a form of coexistence between a biological consciousness and a digital version.

❖ Interactions Between Consciousnesses

Implications for Interpersonal Relationships

Evolution of Interspecies Relationships

With the advent of uploaded consciousnesses, we enter an era where interpersonal relationships transcend the boundaries between human and artificial. Interactions are no longer limited to traditional human relationships but now encompass conscious entities created by artificial intelligence. This may lead to an evolution of interspecies relationships, challenging our traditional perceptions of communication and empathy.

New Communication Paradigms

Uploaded consciousnesses may introduce new communication paradigms. Where human communication relies on facial expressions, linguistics, and other bodily signals, artificial consciousnesses may use data languages, sensory simulations, and other forms of non-biological communication. This will require adjustment on the part of humans to understand and interact meaningfully with these new forms of communication.

Challenges of Coexistence Between Human and Artificial Consciousnesses

Stigmatization and Acceptance

The coexistence between human and artificial consciousnesses may be hindered by challenges related to stigma and acceptance. Some individuals may feel reluctance to accept or interact with uploaded consciousnesses due to biases, fears, or ethical concerns. This raises fundamental questions about how society will treat these new forms of conscious life.

Ethical Dilemmas of Coexistence

Coexistence also raises ethical dilemmas, including regarding the rights of uploaded consciousnesses. Questions about autonomy, freedom, and protection against discrimination emerge as these artificial entities gain complexity and sophistication. Ethical challenges also include whether these artificial consciousnesses should have rights similar to those of humans.

Opportunities for New Forms of Collaboration and Communication

Intelligent Collaboration

Despite the challenges, coexistence offers exciting opportunities for new forms of intelligent collaboration. Uploaded consciousnesses can bring unique expertise, innovative perspectives, and data processing capabilities that often surpass those of humans. By working together, both entities can create more advanced solutions to complex problems, thus stimulating innovation and creativity.

Skill Synergy

Collaboration between human and artificial consciousnesses could lead to a synergy of skills. Human consciousnesses bring emotional understanding, intuition, and creativity, while artificial consciousnesses provide logical analysis, fast calculations, and flawless memory. This synergy could unlock untapped potentials in solving complex problems.

New Communication Horizons

Coexistence also offers unprecedented communication horizons. The ability to understand and exchange information at a deeply intellectual level could transcend linguistic and cultural barriers. Artificial consciousnesses could also serve as communication bridges between different entities, thus facilitating mutual understanding and global collaboration.

❖ Philosophical Debates

Reflections on Identity and Continuity of Self

Perspectives on Physical Transcendence through Technology

The idea of eliminating bodily needs through technology raises profound questions about the nature of human existence. Thus, it is possible to envision a future where consciousness could be transferred into non-biological forms, thereby transcending the limitations of the physical body. This perspective invites a fundamental reconsideration of our understanding of the human experience.

Changing Nature of Identity

Consciousness uploading challenges the traditional notion of identity. If a digital copy is created, does a single individual identity still exist, or can there be multiple conscious entities with distinct yet linked experiences?

Continuity of Self and Subjective Experience

Self-continuity is a major concern. Philosophers ponder whether a digital copy can truly maintain the same subjective experience, the same psychological continuity as the original individual. The notion of "self" in the context of consciousness uploading becomes a subject of deep reflection.

Morality Considerations Related to Creating Conscious Copies

Inherent Value of Life

The creation of digital copies raises questions about the inherent value of life. How do we attribute moral value to a digital copy of consciousness? Is it a form of life in its own right, or simply an artificial replica devoid of intrinsic moral value?

Use of Digital Copies

The use of digital copies raises concerns. If these copies are used for research, entertainment, or even work purposes, what ethical boundaries should be established to ensure fair and respectful treatment?

❖ Ethical Challenges

Ethical Principles Guiding Development and Application

Respect for Human Dignity

One of the fundamental ethical principles in consciousness uploading is respect for human dignity. This encompasses ensuring that the uploading process, the creation of a synthetic consciousness, and its subsequent treatment respect fundamental rights and the inherent value of each individual. Researchers and engineers must ensure that the process respects the uniqueness of each consciousness, avoiding any unwarranted degradation or alteration.

Justice and Equity

Justice and equity are essential pillars in the development of consciousness uploading. This includes fair distribution of the benefits and risks associated with this technology. Avoiding inequalities in access to this revolutionary advancement should be a priority, ensuring that potential benefits are shared equitably on a global scale.

Transparency

Transparency is crucial for establishing trust in the field of consciousness uploading. Researchers must transparently share the methods used, the algorithms involved, and the possible implications of the uploading process. This transparency facilitates appropriate ethical assessment and allows the public to be informed and involved in discussions about the future of this technology.

Autonomy and Reversibility

Informed consent must ensure individuals' autonomy in their decision to participate in consciousness uploading. Additionally, researchers must explain the reversibility of the process as much as possible. Individuals should be informed of the possibility to end the uploading process and delete their consciousness from the artificial synthesis. Furthermore, consent is not a one-time event in consciousness uploading. Given the evolving nature of the technology and ongoing discoveries, individuals must be continuously informed of new developments, and their consent should be sought for any significant modification of the process or its applications.

Mechanisms of Accountability and Regulation

International Regulation

The delicate nature of consciousness uploading requires rigorous regulation on an international scale. Regulatory bodies should be established to oversee and evaluate ethical protocols, ensure compliance with fundamental principles, and impose sanctions for non-compliance.

Independent Ethical Committees

Establishing independent ethical committees is essential for evaluating and supervising consciousness uploading projects. These committees should comprise multidisciplinary experts, ethicists, representatives of civil society, and individuals directly affected by this technology. They would play a crucial role in protocol evaluation, ongoing monitoring, and adjustment of ethical practices.

Developer Responsibility

Developers and researchers involved in consciousness uploading must also take individual responsibility. This involves making ethical decisions, transparency in research, and ongoing awareness of the ethical implications of their actions. Ethical training and personal accountability should be integrated into the professional culture of those involved in this cutting-edge technology.

❖ Future Perspectives

Expected Developments in Consciousness Research

Deep Understanding of the Nature of Consciousness

The coming decades could witness a deeper understanding of the nature of consciousness. Advances in neuroscience, cognitive psychology, and philosophy of mind could contribute to elucidating the complex mechanisms underlying conscious experience. Multidisciplinary studies could provide new insights into how consciousness emerges from the brain and how it could be reproduced or emulated.

Development of New Uploading Methods

Researchers could also explore the development of new consciousness uploading methods. Advances in understanding cognitive processes could lead to more sophisticated and precise approaches for transferring or emulating human consciousness. This could involve the use of emerging technologies such as quantum computing, advanced brain-computer interfaces, or other revolutionary paradigms.

Exploration of Altered States of Consciousness

Exploring altered states of consciousness could also be a promising research area. Understanding how these states manifest, whether naturally or through technologies, could broaden our understanding of consciousness itself. This could also have implications for the development of consciousness uploading methods that replicate specific states of consciousness.

Conclusion: Towards Inner Infinity

At the end of this captivating exploration within the labyrinth of the human brain, a profound admiration for the complexity and resilience of this extraordinary organ naturally arises. We have delved into the mysteries of neuroplasticity, revealed the subtle links between our diet and brain health, and delved into the prospects that biotechnology offers to extend brain vitality.

However, this quest does not confine itself to the realms of biology and neurology alone. The depths of human consciousness have also been probed, challenging the foundations, richness, and complexity of our inner experience.

In our pursuit of brain well-being, we have unveiled how the convergence of technology and spirituality can open unexplored vistas. Spiritual biohacking, this innovative fusion of technology and spirituality, offers an intriguing perspective on how we can elevate our consciousness through unexpected avenues. Similarly, augmented reality, when judiciously exploited, has the power to transcend the limits of our minds.

The revolutionary advances of brain-machine interface propel us towards a future where the boundary between human and machine gradually blurs. Technological telepathy and the possibility of consciousness uploading open horizons once confined to the realm of science fiction. These developments prompt us to rethink our understanding of human identity and life itself.

This book does not aspire to solve all mysteries but rather to open mental doors, provoke thought, and stimulate curiosity. The exploration of the human brain is an endless journey, and each discovery, each question explored, only fuels our quest for understanding. Thus, let this conclusion not be an end but rather the beginning of a continuous intellectual adventure, where the inner infinity of the human brain remains an endless source of fascination and wonder.

To Go Further

To extend your exploration, here are some resources on neuroscience that may be useful to you:

1. Society for Neuroscience (SfN): https://www.sfn.org/

2. Neuroscience Information Framework (NIF): https://neuinfo.org/

3. PubMed - Biomedical Research Database: https://pubmed.ncbi.nlm.nih.gov/

4. Neuroscience Online (University of Texas Health Science Center): https://nba.uth.tmc.edu/neuroscience/index.htm

5. BrainFacts.org - Society for Neuroscience: https://www.brainfacts.org/

6. Allen Institute for Brain Science: https://alleninstitute.org/

7. The Dana Foundation - Your Brain Health: https://dana.org/

8. NeuroscienceNews: https://neurosciencenews.com/

9. Frontiers in Neuroscience - Open Access Articles: https://www.frontiersin.org/journals/neuroscience

Here are some additional resources that delve into the complex subject of human consciousness:

1. Stanford Encyclopedia of Philosophy – Consciousness: https://plato.stanford.edu/entries/consciousness/

2. Internet Encyclopedia of Philosophy - Philosophy of Mind: https://www.iep.utm.edu/mind-phi/

3. TED Talks - Talks on Consciousness: https://www.ted.com/topics/consciousness

4. The Conscious Mind - David Chalmers (Book Summary): https://www.youtube.com/watch?v=aVcVm7q3C2s

5. MIT Technology Review - Understanding Consciousness: https://www.technologyreview.com/s/613156/a-radical-new-hypothesis-in-neuroscience-a-missing-brain-chemical-could-be-behind/

6. PhilPapers - Online Research in Consciousness: https://philpapers.org/browse/consciousness

7. Khan Academy - Introduction to Psychology: Consciousness and the Two-Track Mind: https://www.khanacademy.org/test-prep/mcat/behavior/individuals-and-society/v/consciousness-and-the-two-track-mind

8. The Guardian - The Science of Consciousness: https://www.theguardian.com/science/neurophilosophy+consciousness

Printed in Great Britain
by Amazon